TOTU

(Thoughts on the Universe)

Ian Atkinson

First Edition

Dedication.

This book is dedicated to all life on Earth.

I would like to thank my wife, Caroline, my children, my parents and all my ancestors for everything that they have done to make this book possible.

I would also like to thank all the scientists and engineers that have contributed to this book. Firstly, I would like to thank Isaac Newton for guiding me. I would then like to thank Michael Faraday and James Clerk Maxwell for their thoughts on electricity, magnetism and light. I would like to thank Charles Darwin for his ideas on evolution which have shaped my thoughts. I would also like to thank all the staff and personnel of NASA not only for landing men on the Moon but also for placing a laser reflector on the Moon and, more importantly, making the data from those experiments available to all.

Finally I would like to thank the people and governments of England, South Africa and the United States of America for everything that they have done for my family and I over the years.

Introduction.

I was first introduced to Newton's laws of motion in February 1964 when I started university as a student in chemical engineering. One question in particular intrigued me. What causes gravity?

I also became increasingly interested in the question of what causes the phenomena of electrical charges and magnetism. I was fascinated by light. Fairly early on in my life I acquired a copy of Newton's Optiks.

During my early twenties I first started forming the ideas which I will discuss in this book. In 1969 I formed my first tentative ideas about gravity as well as my ideas on a possible way of rowing through the aether. I do remember feeling at that time that Newton's idea on light being transmitted as corpuscles must be correct.

During those days I also read Michael Faraday's book on his studies in electricity and magnetism as well as Maxwell's book, in two volumes, on electricity and magnetism. I must admit that at that time I could not fully follow what Maxwell was talking about. Nearly forty years later, I have come to realize that one of the great tragedies in science was that Maxwell died of illness at a relatively young age before he had time, in what should have been his later years, to clarify what he meant for younger readers.

I was over forty years old before I first read Newton's Principia. His elegant treatment of problems that I had been wrestling with impressed me. What particularly impressed me was that Newton was able to prove all of his proofs using geometry. During the last eighteen years, since reading Newton's Principia, I have gradually been refining my ideas on the universe.

The one question that always eluded me over the years was what was the aether. Both Maxwell and Faraday were convinced that an aether had to exist. One point in Faraday's writings has stuck with me over the years. It was Faraday's belief that electrical energy is stored in the aether itself. Over the last forty years I have also come to the conclusion that the aether exists and holds tremendous energy. Finally, on the morning of my fifty-seventh birthday, on December 23, 2003, as I was waking up I finally had an explanation that satisfied me as to what the aether could be.

I have made a conscious decision, in writing down my thoughts, to try to write them down as simply as possible, using as little mathematics as possible, in such a way that a young person with very little formal training can follow my thoughts. I have also, where possible, followed Newton's methods in attempting to prove things using geometry rather than using calculus for several reasons. Firstly it forces one to think about what one is dealing with rather than concentrating on the mathematical manipulation of equations. Secondly, my feeling is that when dealing with geometry, any assumptions that one makes are more clearly understood than when dealing with calculus. And finally, when dealing with particles in motion, either in straight lines or revolving about each other, one is dealing with geometric relationships between bodies that are, I feel, more easily visualized using geometry rather than equations.

Persons such as Newton, Faraday and Maxwell who can see the problems in the existing theories of the day are quite rare. As our theories become more complicated, the task of discovering any faults in our theories become more complicated, particularly if our theories are presented in complicated ways that require years of training in mathematics to understand the theories. I am convinced that young fresh minds have more chance of finding errors in our theories than do older minds which are more resistant to new ideas.

It may be hundreds of years before a young mind can see through the faults of our present day theories and extend our knowledge by suggesting the correct explanation for the phenomena of electricity, magnetism and light. When that mind is ready to start working on the problem, I hope that my thoughts on the matter will be as much of a help to that person as Newton's, Faraday's and Maxwell's have been to me. And I encourage that person to point out any problems in my theories or any assumptions that I have made in the same way that I have pointed out implicit assumptions that Newton made in his Principia.

I am sure that Newton, if he were alive today, would be the first to accept and understand the need to clarify assumptions that he made in his Principia as well as to correct some mistakes that he made in his thinking on light, particularly his corpuscular theory. In fact, I am quite sure that if Newton were alive today and had the benefit of all the scientific knowledge that has been accumulated in the last three hundred years since his time, then not only would he be far more capable than me in extending his original work on gravity but he would also, I am sure, be able not only to point out assumptions which he originally made but also errors that I have made both in my understanding of his original work as well as in my own theories. So, in the same way, do I accept the need for those that follow me to clarify assumptions and, where necessary, correct errors which I have made and I encourage them to do so.

Newton, in his General Scholium at the end of his Principia, mentioned that science must be based upon experiments and experimental verification of theories. Newton was very careful, throughout his Principia, to limit himself to what could be experimentally verified in his time. Newton had experienced, first hand, the difficulties that were encountered when wrong assumptions about the sun and planets had been accepted for thousands of years. We tend to forget that in Galileo and then Newton's time the Earth was still widely considered to be the center of the universe and that theories of vortices abounded to explain the movement of the planets about the Earth. It is not surprising, therefore, that Newton felt very strongly that theories had no place in science unless they could be supported by experimental verification.

Newton might have been very surprised to learn that from his time until now, a span of nearly three hundred years, virtually no advances have been made in our understanding about gravity. The only real extension to our knowledge about gravity since Newton's time is that we have found that the path of light can be bent by gravity.

One of the problems that face all scientists is that they have rather short life spans when compared to the times taken for astronomical processes to take place. Even something as simple as Haley's comet only orbits the sun once every seventy five years or so. The practical effect of this was that Haley and Newton were both dead before their theories and calculations about Haley's comet could be verified.

When one considers that it takes light 80,000 years to merely cross our galaxy and that it takes our sun perhaps 240 million years to travel around our galaxy once, once should realize that some scientific experiments that might be needed to verify a theory might take hundreds or thousands of lifetimes to complete.

One is then left to ask the question as to how can science proceed in the long run if it is impossible for a scientist to verify his theory during his lifetime. Should the scientist, as Newton to some extent did, keep quiet and not tell anyone about his ideas. In some ways this might be a good idea. There has most definitely been a tremendous amount of controversy since Newton's time about his corpuscular theory of light.

But in other ways it might be a bad idea. For if a scientist makes a correct guess about how something happens and then never tells anyone about it, due to his inability to verify his guess experimentally, then the progress of science is hindered until someone else makes the same guess again. And when the same guess is made again by a second person in some later generation, if it is still impossible to verify the guess during that second person's life span, then science can never progress until a more cooperative system is set up whereby people in subsequent generations can either verify or disprove the guess. And if a scientist makes a wrong guess, as Newton did with his corpuscular theory of light, then even if it cannot be verified or disproved during his lifetime at least other scientists in the years following the scientist's death will be able to disprove the incorrect parts of the theories.

I feel, therefore, that Einstein's approach of putting forward a theory and leaving it to subsequent generations to verify it has the benefit of at least giving other scientists the opportunity to know about the theory and to either verify or disprove the theory if and when it becomes possible to do so. The only proviso that I would put to this method of progress is that one must be very careful not to fall into the trap of believing, just because some famous scientist had a theory or that just because a theory has not been proved to be incorrect for a long time, that the theory is therefore correct. This is a trap that many fell into during Newton's lifetime in persisting with theories of vortices and theories that the Earth was the center of the universe that had been around since the time of the ancient Greeks and probably before that. No matter how famous the scientist or how elegant the theory or how long the theory has been around for, if it can be proved experimentally that the theory, or any part of the theory is wrong then the scientist was, at least in some respects, dead wrong in his guess and the theory is incorrect and should be discarded or modified without delay as soon as it can be shown to be incorrect.

I personally feel very grateful to Newton for adding his General Scholium at the end of his Principia in which he summarized his thoughts of a lifetime on gravity. His thoughts on gravity have been more useful to me than the ideas of all other persons combined that I have ever encountered. For Newton was wrestling in solving the very same problem that I have been working on for the last forty years and his thoughts at end of the Principia have guided me upon the path to my theories.

As I mentioned, when I was young I felt that Newton's corpuscular theory of light must be correct. Subsequent knowledge as well as ideas and writing by Maxwell have convinced me that Newton was wrong on his corpuscular theory of light. But it has been better for science that Newton did advance his theory even if it proved to be wrong. For if science is to evolve, it will be ultimately be, as it always has been, the survival of the fittest theories.

I will split what I wish to say into five main sections.

The first section will be to clarify assumptions that Newton made in his Principia and to extend Newton's concepts to properly take into account what would happen if gravity moves at a finite speed. I will then demonstrate how to determine the speed of gravity using scientifically measured quantities and will obtain a rough estimate of the speed of gravity. I will also use the concepts used to estimate the speed of gravity to determine the maximum time that the Earth and Moon have been moving apart

In the second section I will present my theories on energy and matter and propose a theory to explain what the aether is. I will then, using this model of the aether, explain a mechanism by which gravity may be explained. I will then, using this theory, derive Newton's law of gravitation from first principles. I will show that Newton's law of gravitation is in fact a special case of a more general law and that Newton's law of gravitation only works under certain conditions. I will also then explain how Newton's law of gravitation must be modified when dealing with massive bodies, such as black holes. Finally I will show, based upon my theory of gravity, that it is not always valid to apply Newton's third law of motion when considering problems concerning gravity.

In the third section I will present some of my theories on the universe in general. I will try to present an alternative to the big bang theory and the theory of the expanding universe. I will also try to explain why galaxies and our solar system have the shapes that they do. I will give some very rough ideas of mine concerning electricity and magnetism. I know that I do not understand the underlying principles of electricity and magnetism. I would like to use some of the ideas developed in the first section on gravity and apply them to electricity and magnetism as I feel that some of these ideas might be useful in understanding quantum theory as well as better understanding the action of bodies that move quickly with respect to the speed at which electricity and magnetism are propagated.

In the fourth section I would like to present an interesting concept that I thought of in 1969, shortly after the first men landed upon the Moon, for a method of propulsion that might essentially enable us to "row" through the aether and make space travel a viable proposition.

In the fifth section I would like to present some simple thoughts on measuring and constructing angles and propose a new system of measuring angles as well as longitude.

I am rather reluctant to publish this work for several reasons. The first is that what little that I know about science and the universe in general has been largely self taught. This has resulted in vast amounts of knowledge which are available and which are probably very relevant to the subjects which I have discussed being totally unknown to me. The second is that there are great gaps in my mathematical abilities. I do not really understand all the mathematics involved in Maxwell's theory of electricity and magnetism. I wish that I did. I cannot help feeling that if Maxwell had had the benefits of reading my theories on energy and the aether that he would then have been able to give a much more complete description of electricity and magnetism. The third is that, as a result of my deficiencies mentioned above, I am rather afraid that I will make a fool of myself by publishing my theories.

On the other hand, I feel that my concepts about matter and energy, which are based on easily understood concepts, make more sense to me than the alternatives which use mathematical equations as the starting point of the theories. My core belief is that if the basic principles of a theory cannot be explained simply in terms that an average person can understand then there is a problem with the theory. For there are two different phases involved in any theory. The first is to explain the fundamentals in simple terms which the average layman can understand. The second is to derive mathematical models, based upon the fundamental ideas, to be able to predict exactly what will happen under a given set of conditions. Thinks about how science has developed from Newton's time. Newton gave us four very fundamental laws, which he explained simply in words, to explain how gravity and forces and movements could be calculated. Since then a large and very complicated body of mathematics has been developed to enable scientists and engineers to apply these simple fundamental laws to a wide range of problems in science and engineering. I am sure that many who are more capable than I am will be able to develop the mathematics required to solve some of the problems that can be solved once my theories are understood.

My main purpose in publishing this work is not to help those who will apply my theories to solve problems, Rather, the purpose of this work is to help that one mind who will follow me and extend my theories to explain the fundamental principles for electrical charges, magnetism and light. I hope that this work will help that person in the same way that Newton, Faraday and Maxwell have helped me.

Ian Atkinson
October 2006
Connecticut, USA

Section 1 - To determine the speed to gravity.

Introduction

I will propose a theory, in Section 2, to explain the cause of gravity and to derive the equation for Newton's law of gravity from first principles.

In this section, however, I will concentrate on a slightly different question, namely the speed at which gravity travels. My theory to explain the reason for gravity may or may not be correct. I feel that as yet there is insufficient experimental evidence to support my theory as to the cause of gravity. If gravity does move with a finite speed, however, then one can still carry on a separate investigation to determine the speed of gravity irrespective of whether of not my theory for the cause of gravity is correct .

In terms of determining the speed of gravity, I have attempted to present this section in such a way that this section can be considered totally separately from the next section, irrespective of whether or not one believes my theory in the next section. What is presented in this section is really independent of the propositions in the next section.

This section might seem a little strange and awkward for the reader that understands and agrees with what I say in Section 2. In some cases I almost seem to be repeating some arguments that I make in Section 2. This is unavoidable, however, as this section is designed to demonstrate, from first principles, how the speed of gravity might be determined irrespective of whether or not the reader believes what I have said in Section 2.

Laws of transmission of force.

Law 1.
Suppose a first body is acted upon by an external force originating from, or caused by, a second body and this force

requires a finite time to be transmitted between the two bodies. Suppose that at any instant of time, t, the force acting upon the first body had required a time dt to travel from the second body to the first body. Then at this instant of time t the force acting upon the first body will be identical to the force that would result if the force was transmitted instantaneously, i.e. with infinite speed, and the second body was, at time t, at the position and state that it had been at the earlier time t-dt.

Law 2.

If two bodies act upon each other with a mutual type of force, such as gravity, then each body is acted upon independently by the other body by the force which originated from, or was caused by, the other body at the position that the other body was when the force left the other body on its way to the that body.

Example

Suppose we look at a plane flying quickly through the air. Although we can see where the plane is, the force of sound of the plane arriving at our ears at any instant of time is that of sound which left the plane at the position that the plane was at a time interval earlier, that time interval being the time that it took for sound to reach us from the earlier position to our present position. For this reason, the plane sounds to us as if it is in an earlier position than it actually is.

Suppose that we could not detect light and could not see where the plane was. If we were to rely solely upon our ears for detecting the position of the plane, we could never be sure of where the plane was at any instant of time. The best that we could then do would be to determine where the plane was at an earlier time when the sound left the plane on its way to us.

In most situations, if we could not detect light our ears would still give us a very good account of where most objects are. It is only when distances are great or objects are moving at high rates of speed with respect to the speed of sound, so that that it takes a measurable time for sound to reach our ears, that we would have to take into account the speed of sound in trying to determine positions accurately.

In a similar way, if any action takes time to be transmitted from one place to another, then the best that we could do in determining a position of the source of such an action, if we had to rely upon the action itself for information about the source, would be to determine where the source was at the time when the action left the source as opposed to the later time when that action reached us. And in a similar way, it would only be when distances are very great or objects are moving at high rates of speed with respect to the speed at which the action is being transmitted, that we would have to take into account the speed at which the action is being transmitted.

Newton's Law of Gravity

Newton, in his Principia, proposed that the force of gravity acting between two bodies is proportional to the product of the masses of the two bodies and inversely proportional to the square of the distance between the bodies.

Newton's law of gravity is commonly expressed in the form:

$$F = \frac{G.M_1 M_2}{R^2}$$

Newton, in his Principia, was the first person to explain that all bodies are acted upon by the force of gravity and that it is this force of gravity that keeps the planets in their orbits about the sun. We take this for granted today. In Newton's time this was not so, however, and Newton's main job was one of showing that the motion of the Earth, Moon and planets about the sun could be explained by the concept of gravity. Newton faced tremendous obstacles and had to invent the calculus in order to prove his theory. It is hardly surprising that Newton made some simplifying assumptions in his wonderful geometric proofs and reasoning concerning gravity. These proved to be quite adequate in explaining all the phenomena which could be measured in Newton's time.

Newton was also well aware that he did not understand what caused gravity and, given the knowledge of his day, he could not progress further than he did. His thoughts in his General Scholium at the end of his Principia, given what was known in his day, are fascinating and are responsible for helping me to solve some further questions concerning gravity.

The most important assumption that Newton made throughout his Principia, either knowingly or not, was that gravity is transmitted instantaneously between any two bodies.

I question this by simply asking how can gravity act instantaneously? How can one assume that gravity can pass between two bodies at the furthest ends of our known universe instantaneously. Surely there is some physical mechanism by which the force of gravity is transmitted between the bodies. And if there is any physical mechanism by which gravity is transmitted, how can it act instantaneously? And if it does not act instantaneously, then as I will illustrate in what follows, if the bodies are not stationary and are far enough apart, there must always be slight differences in the directions between the forces acting on the two bodies, i.e. they cannot act on the line joining the bodies.

This brings into question the second assumption which Newton made in his Principia concerning gravity. Newton assumed that his third law of motion could be applied directly when considering gravity.

Newton's Third Law of Motion.

Newton, in his Principia, states his third law of motion as follows:

"To every action there is always opposed an equal reaction or, the mutual action of two bodies upon each other are always equal, and directed to contrary parts."

Newton, in his Principia, after stating his third law of motion gave examples in which he discussed pressing a stone with a finger and a horse pulling a rope. Newton, in defining his third law, assumed that the time taken for the forces to be transmitted between the two bodies is negligible compared to any possible change of position between the bodies during that time interval that it takes the force to be transmitted between the two bodies. In practice, these effects occur so quickly and distances involved are so small that we can normally ignore them from our considerations.

Newton assumed that his third law was true in all his geometric proofs in his Principia concerning gravity and actions at a distance. While a very, very good approximation, this is not entirely true. In all analyses concerning gravity or other actions at a distance such as magnetism or electrical charges, we must always take into account the time taken for the force in question to be transmitted from one point to another. When we consider our solar system, the distances are so small and speeds of the sun and planets are so slow in comparison to the speed of gravity that the errors caused by assuming, as Newton did, that gravity acts instantaneously are so small as to be almost impossible to measure.

Newton, in his time, was not aware of galaxies or clusters of galaxies or the massive distances involved when considering these bodies. Light can pass from our sun to the Earth in eight minutes but requires eighty thousand years to travel across our galaxy and billions of years to reach us from the furthest observed galaxies. It is quite conceivable and, as I will demonstrate later, quite probable that gravity actually does require a measurable time to pass across our galaxy. When considering the motion of galaxies, the distances are so great that one must take into account the effects caused by the time taken for gravity to pass from one side of the galaxy to the other.

And when one does take into account these effects, one must always be careful not to incorrectly assume that Newton's third law can be applied directly to the problem in question.

For Newton's third law is really only true if the time taken for the transmission of the action and the reaction is negligible and can be neglected.

One must, therefore, be very careful before using Newton's third law directly when dealing with those forces commonly referred to as actions at distances, namely gravity, electricity and magnetism. There could well be situations where the time taken for a force to travel from the one body to the other cannot be neglected. Under such circumstances an analysis must be done taking into account the time that forces take to travel between the bodies. In general, one may then neither assume that the forces act in the same direction or are even equal to each other.

Furthermore, one may not be even be able to use formulas derived by Newton in his Principia. For in all of the geometric proofs throughout Newton's Principia, Newton implicitly assumes that his third law holds true for actions at a distance by assuming that that forces are transmitted instantaneously and that any force between two bodies act along the straight line between the two bodies.

Once one understands the assumptions that Newton made, either knowingly or not, in his Principia one can still apply most of Newton's logic to problems concerning gravity. In particular, all the work that Newton did in showing what type of forces would be required to make bodies move in certain geometric paths is valid. For example, Newton showed that in order to keep a body moving at a constant velocity around a path being the circumference of a circle, a force is required to continually modify the direction in which the body moves. This resultant force must act at all times in the direction of the line from the body to the center of the circle and is proportional to the square of the velocity of the body and inversely proportional to the distance of the body from the center of the circle. Newton's analysis in this case is correct irrespective of the speed at which the force is transmitted. This is due to the fact that, at any instant of time, we are analyzing the force that actually arrives at the body at that instant of time, irrespective of where it came from. And at the instant of time when the force modifies the direction of the body, it has already reached the body and there is thus no time delay involved between the action of the force upon the body and the change of direction of the body.

Questions to ask about gravity.

The question to ask, therefore, is what is the speed of gravity and how can it be measured. What is clear is that Newton's law of gravity certainly is an extremely good approximation for measuring the effects between the sun, planets and all their moons that have been observed to date. This must mean that gravity moves so quickly that the time taken for gravity to move between the sun and all the planets and the planets and their moons must be so small as to be virtually impossible to measure. For the faster gravity moves, the less noticeable differences due to the time taken to move between any two bodies will be.

In following my attempts to calculate the speed of gravity, I would like to ask the reader to follow my arguments with an open mind. Please put aside all your preconceived notions of gravity and everything that you have been taught to date that has been based on the assumption that gravity moves instantaneously. Try instead to think about what happens when we perhaps follow the more logical path of assuming that gravity does not act instantaneously.

Modification to Newton's Law of gravity to take into account the speed at which gravity moves.

Suppose two bodies, A and B, with masses M_A and M_B interact with each other due to the force of gravity between the two bodies. Suppose further that gravity does not act instantaneously between the bodies but rather takes a finite time to travel between the bodies. Suppose further that the two bodies, A and B, are moving relative to each other with respect to a non rotating frame of reference. Under such circumstances, during the finite time that it takes gravity to pass between the bodies, each body will have moved with respect to the other body.

My feeling is that, under such circumstances, at any instant of time each body will be acted upon by gravity from the other body from the position that the other body was at a time earlier to the current time by the time that it took for the gravity to pass from the other body to that body.

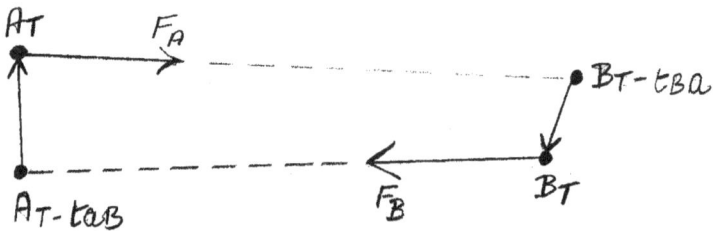

Section 1 - Figure 1

Suppose at an instant of time, T, the two bodies A and B are at positions A_T and B_T and have masses M_A and M_B respectively. Suppose that gravity is transmitted at a finite speed, say V_g. If bodies A and B are moving relative to each other, then at time T, body A at position A_T cannot be aware that body B has moved to position B_T. For the gravity which does leave body B when it is at position B_T, leaves at time T and would then require a certain time after that to reach body A. But by the time that it does reach the position A_T where body A was at time T, body A will have moved from point A_T to a different position. In addition, body B will have moved from the position at B_T to a different position. The gravity reaching body A at time T must, in fact, have spent a certain time in traveling from B to A. Suppose at the instant of time T the gravity reaching body A at position A_T had taken a time interval tba to travel from B to A. The gravity reaching A from B will, therefore, have left body B at a time tba before time T and from the position that B was at that time, $T - tba$. Let us suppose that the position of B at this earlier time was B_{T-tba}. Therefore the distance B_{T-tba} to A_T, is thus equal to V_g times tba.

If we consider body A at position A_T it is being acted upon by gravity from body B at position B_{T-tba}. The force of gravity on body A from body B would be given by the formula

$$F_A = \frac{G.M_A.M_B}{(\text{distance from } B_{T-tba} \text{ to } A_T)^2}$$

And the direction of the force of gravity acting upon body A at time T will be in the direction of the straight line from A_T to B_{T-tba}.

In a similar manner, if tab is the time that it took gravity to travel from A to arrive at body B at time T and position B_T, then distance A_{T-tab} to B_T equals V_g times *tab*. The force of gravity acting upon body B at time T will thus be:

$$F_B = \frac{G.M_A.M_B}{(\text{distance from A}_{T-tab} \text{ to B}_T)^2}$$

And the direction of the force of gravity acting upon body B at time T will be in the direction of the straight line from B_T to A_{T-tab}.

If we were to assume that gravity moved with infinite speed, then the time taken for gravity to move between A and B would be zero and we would have

$$tba = tab = 0$$
$$A_{T-tab} = A_T$$
$$B_{T-tba} = B_T$$

Substituting these values into the above two equations for F_A and F_B we would get

$$F_A = F_B = \frac{G.M_A.M_B}{(\text{distance from } A_T \text{ to } B_T)^2}$$

The two equations would thus simplify into Newton's equation for gravitation where the speed of gravity is assumed to be infinite. Both forces would then be equal and act in opposite directions along the straight line joining A_T and B_T.

If, however, gravity does not move with infinite speed then the more accurate equations,

$$F_A = \frac{G.M_A.M_B}{(\text{distance from } B_{T-tba} \text{ to } A_T)^2}$$

and

$$F_B = \frac{G.M_A.M_B}{(\text{distance from } A_{T-tab} \text{ to } B_T)^2}$$

should be used to more accurately take into account the speed of gravity. Please note that in the above equations time *tba* need not necessarily be the same as time *tab*. Hence F_A need not necessarily be equal to F_B. And also note that the direction of force F_A need not necessarily be in the exact opposite direction to force F_B. This will be illustrated in the following rather extreme example.

A numerical example illustrating effects caused by the speed at which a force is transmitted.

Let us consider a very simple example to illustrate the effects caused by the speed at which a force moves. Suppose we have two bodies, *A* and *B* moving in directions perpendicular to each other as shown on the following diagram and acted upon by a force, such as gravity, which acts mutually upon both bodies. Let us further assume that the distances involved are so great and the velocities of *A* and *B* are so high that the effect of the force between *A* and *B* is so small that it has a negligible effect in changing the directions or velocities of *A* and *B* .

Suppose body *A* moves along the X axis in a positive direction (from left to right) with a constant velocity of 2 units of distance per unit of time. Suppose that at time 0 body *A* is at a position of -4 on the X axis. Then at any time *T* the position of body *A* on the X axis will be given by the formula

$$x_T = 2T - 4$$

Suppose that body B moves along the Y axis in a positive direction (from bottom to top) with a constant velocity of 1 unit of distance per unit of time. Suppose that at time 0 body B is at a position of -10 on the Y axis. Then at any time T the position of body B on the Y axis will be given by

$$y_T = T - 10$$

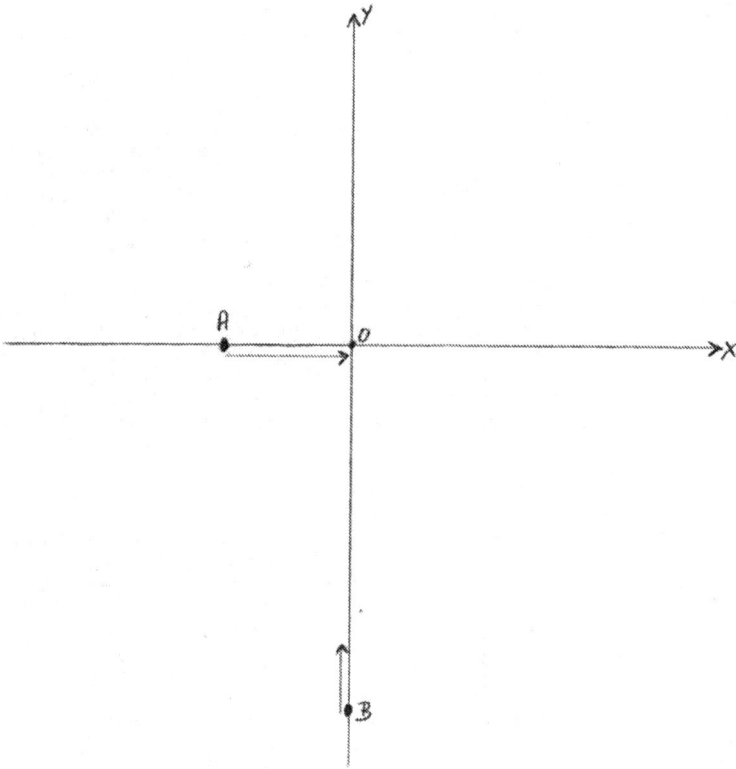

Section 1 Figure 2

Let us further assume that the force in question moves with a constant velocity of 10 units of distance in one unit of time. Hence the distance moved by the force in any interval of time, t, is given by the formula:

$$\text{distance} = 10t$$

Now let us derive the formula to find, at any time T, the position where body B was at the time the force left B on it's way to body A. Suppose that at the time T that body A receives the force from B. Then that force required a time t to travel between the original position of B at time $T - t$ and the final position of body A at time T. So the force left B at time $T - t$ and arrived at body A at time T.

So at time T, the position of A on the X axis when it received the force from B is given by

$$x_T = 2T - 4$$

The position of B, on the Y axis, at the earlier time $T - t$, when the force left B on it's way to meet A at time T, is given by:

$$y_{T-t} = (T - t) - 10$$

Now the distance that the force traveled on it's way from B to A is the distance between where A is at time T and where B was at time $T - t$. Due to the very simple nature of this example where A and B move along lines perpendicular to each other, this distance is easily determined by Pythagoras's theorem as follows:

$$\text{distance} = \sqrt{x_T^2 + y_{T-t}^2}$$

but

$$\text{distance} = 10t$$

hence

$$\text{distance}^2 = (10t)^2 = x_T^2 + y_{T-t}^2$$

Substituting in the values for x_T and y_{T-t} into the above equation gives:

$$100t^2 = (2T - 4)^2 + (T - t - 10)^2$$

or

$$100t^2 = (2T - 4)^2 + ((T - 10) - t)^2$$

or
$$100t^2 = (2T-4)^2 + (T-10)^2 - 2t(T-10) + t^2$$
or simplified:
$$99t^2 + 2(T-10)t - (2T-4)^2 - (T-10)^2 = 0$$

This is a simple quadratic equation in t^2 which can easily be solved for t for any given value of T. Once the value of t is found for any value of T, the position of B can be found at the earlier time $T-t$ when the force left B on its way to A by simply substituting t into the equation
$$y_{T-t} = (T-t) - 10$$

In a similar manner, the formula to find, at any time T, the position where body A was at the time the force left A on it's way to body B can be found. In this case, if t were the time taken to get from A to B to arrive at B at time T then
$$100t^2 = (T-10)^2 + (2(T-t)-4)^2$$
or
$$100t^2 = (T-10)^2 + ((2T-4)-2t)^2$$
or
$$100t^2 = (T-10)^2 + (2T-4)^2 - 2.(2T-4).2t + 4t^2$$
or simplified:
$$96t^2 + 4(2T-4)t - (T-10)^2 - (2T-4)^2 = 0$$

Once again, the value of t can be found for any given value of T. The position of A can then be found at the earlier time $T-t$ when the force left A on it's way to B by substituting this value of t into the equation
$$x_{T-t} = 2(T-t) - 4$$

The following table, marked as Section 1 – Table 1, illustrates for each value of T from 0 to 20 the actual positions of A and B on the X and Y axis in columns 2 and 3 respectively. Column 4 indicates the time required to travel from B to A to arrive at A

at time T. Column 5 indicates the position of B on the Y axis at this earlier time when the force left B to go to A to arrive at A at time T. Column 6 indicates the time required to travel from A to B to arrive at B at time T. Column 7 indicates the position of A on the X axis at this earlier time when the force left A to go to B to arrive at B at time T.

1	2	3	4	5	6	7
Time T	Posn of A On X axis at time T	Posn of B on Y axis at time T	Time from B to A to arrive at A at time T	Position of B on Y axis when force left B to arrive at A at time T	Time from A to B to arrive at B at time T	Position of A on X axis when force left A to arrive at B at time T
0	-4	-10	1.188	-11.188	1.186	-6.371
1	-2	-9	1.022	-10.022	0.984	-3.967
2	0	-8	0.889	-8.889	0.816	-1.633
3	2	-7	0.806	-7.806	0.703	0.595
4	4	-6	0.788	-6.788	0.657	2.685
5	6	-5	0.837	-5.837	0.682	4.636
6	8	-4	0.940	-4.940	0.761	6.477
7	10	-3	1.080	-4.080	0.877	8.245
8	12	-2	1.243	-3.243	1.017	9.967
9	14	-1	1.421	-2.421	1.170	11.660
10	16	0	1.608	-1.608	1.333	13.333
11	18	1	1.802	-0.802	1.503	14.994
12	20	2	2.000	0.000	1.677	16.647
13	22	3	2.201	0.799	1.854	18.293
14	24	4	2.405	1.595	2.033	19.934
15	26	5	2.611	2.389	2.214	21.571
16	28	6	2.818	3.182	2.397	23.206
17	30	7	3.026	3.974	2.581	24.839
18	32	8	3.235	4.765	2.765	26.470
19	34	9	3.445	5.555	2.951	28.099
20	36	10	3.655	6.345	3.136	29.727

Section 1 – Table 1

As one can see, by examining columns 4 and 6, the time taken for the forces to arrive at the two bodies at any instant of time are different from each other. In this numerical example, the distances traveled are just ten times the time taken. Hence the distances traveled by the force arriving at the two bodies at any instant of time are also different. Hence the magnitudes of the force from B acting on A at any instant of time would be slightly different from the magnitude of the force from A acting on B as these forces are inversely proportional to the squares of the distances involved. I will leave it to the reader to verify that the directions of the forces acting on A from B is different from the direction of the force acting on B from A at any instant of time.

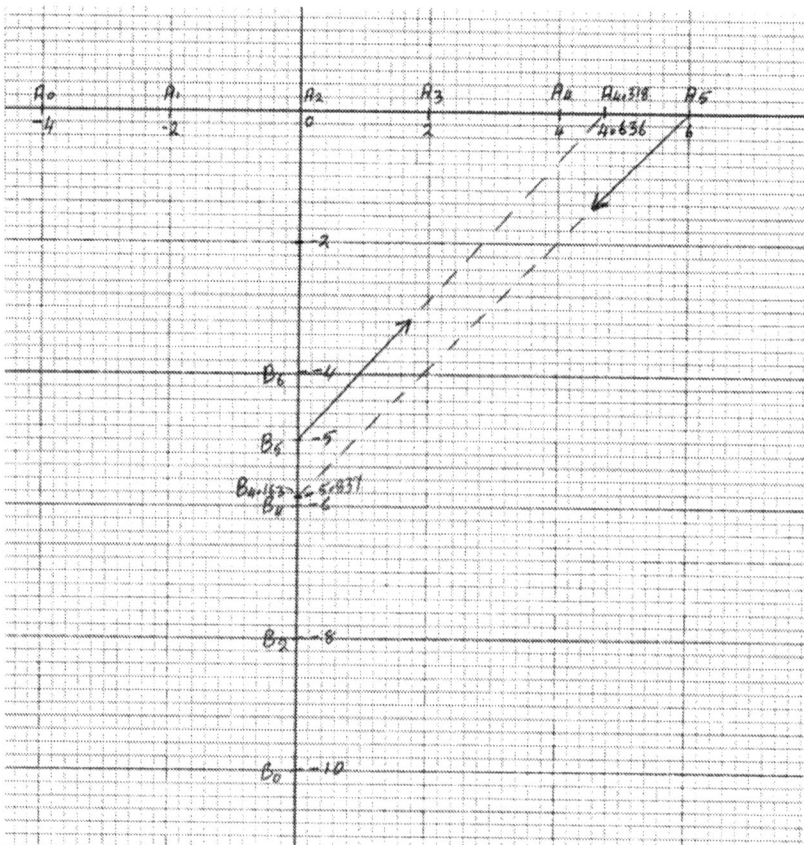

Section 1 – Figure 3

In the figure shown above, marked as Section 1 Figure 3, I have indicated the position of A at time $T = 5$. From Table 1 earlier it may be found that gravity took 0.837 to arrive at A at time $T = 5$. Hence gravity left B at time 4.163 to arrive at A at time $T = 5$. I have also indicated the position of B at time $T = 4.163$, which is at point −5.837 on the Y axis and is where B would have to have been for the force to arrive from B to A at time $T = 5$. I have also indicated the position of B at time $T = 5$. From Table 1 column 6 the force would have required a time of 0.682 to travel from A to B to arrive at B at time $T = 5$. From Table 1 column 7 the position of A at time $T = 4.318$ would have been at point 4.636 on the X axis. As can be seen from the above example, the direction of the force from A to B is different to that from B to A at time $T = 5$. In addition, since the time take by the two forces in arriving at time $T = 5$ are different, the distances that they traveled are different and the magnitudes of the two forces are different.

In this example, we assumed that the two bodies traveled in perpendicular straight lines and that their motions were not affected by each other. Even with these simplifying assumptions, the mathematics required to solve the positions at any time are slightly complicated. If one were to now assume, as happens in practice, that the forces acting between the two bodies affected the motions of each of the bodies so that they no longer moved along the X and Y axes, then mathematics required becomes much more difficult.

My purpose at present is to merely illustrate what could happen when dealing with bodies which interact with each other in situations where the speed of bodies is significant in relation to the speed at which forces are transmitted. The most important thing that I hope that this example will illustrate to the reader is that in a situation where an apparently mutual force acts between two bodies, such as gravity, if the speed of the bodies is significant compared to the speed at which the force between the bodies is transmitted, then at any instant of time the "mutual" forces acting on the two bodies are neither equal nor do they lie in the exact opposite direction.

What I really want to concentrate on at present is what happens in the case of gravity. It appears to me as if the speed of gravity is so high that for all practical purposes, the speed of bodies moving away from each other relative to the speed of gravity is very, very small.

Is it possible for two bodies orbiting about each other to maintain unchanging stable orbits if gravity moves with a finite speed?

Let us consider, as an exercise, the case of two spherical bodies, A and B, of identical masses M, initially rotating about each other in perfectly circular stable orbits and held in those orbits by gravity.

As is well known, under classical Newton mechanics, the two bodies will revolve about their common center of gravity. In the case where the two bodies are of exactly equal masses, this common center of gravity will be the exact center of the line joining the centers of the two bodies. Let us now assume that it takes a finite interval of time I_g for gravity to move the distance between the two bodies.

Suppose that the two bodies, are initially at points A_0 and B_0 equidistant from their common center of gravity O. Suppose further that after time I_g the bodies have moved to points A_1 and B_1 on the circle as shown in Section 1 – Figure 4 below. Now at this time body A, at position A_1 will be pulled by body B when body B was at position B_0. Similarly, body B, at position B_1, will be pulled by body A when body A was at position A_0. Let us call this angle between the line joining the actual positions of A and B at any instant of time and the direction of the gravitational force on A and B the angle of gravitational deviation, α. Then

$$\alpha = \angle A_0 B_1 A_1 = \angle B_0 A_1 B_1$$

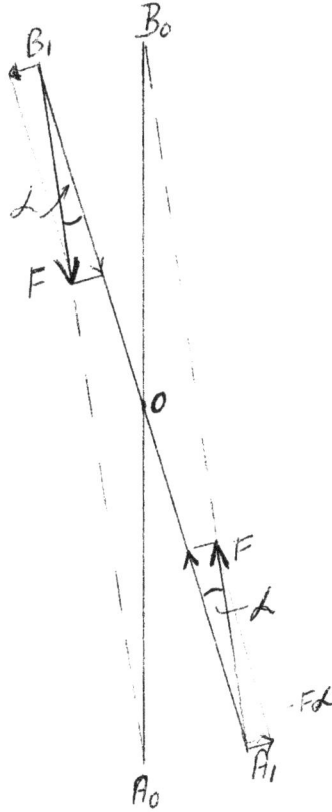

Section 1 - Figure 4

One thing is clear in the example chosen. However bodies A and B move, due to the fact that their masses are assumed to be identical, there must be a symmetry of movement about their common center of gravity O. For any action that would impel body A to move in any way from its circular orbit there would be an identical, mirror image, action upon B that would impel body B to move in an equal but opposite direction to that of body A so that the point O would always remain on the exact center of the line drawn between A and B at any time. Hence, if the two equal bodies A and B rotate about each other in any sort of spiral or otherwise unstable orbit, whatever the paths of A and B might be, the point O would always remain as the common center of gravity. So for the purposes of this example, we could regard point O as an immovable center of gravity.

The second thing that is clear is that the shorter the time required for gravity to pass between A and B, the closer the orbits of A and B will remain to the original circular orbit. For if gravity moved instantaneously, i.e. with infinite speed, between bodies A and B then the bodies would remain in their original circular orbit.

The third thing that is clear is that the longer it takes for gravity to pass between A and B relative to the time that it takes bodies A and B to rotate once about their common center of gravity, the more distorted the orbits will be from the original orbit.

Now we know from experiment that the orbits of planets are very nearly those predicted using Newton's methods. This means that, in practice, the time taken for gravity to move between the sun and all its planets must be very short compared to the time taken for planets to circle the sun.

So let us further assume that the time, I_g, required for gravity to move between A and B is extremely small compared to the time taken for A and B to complete one complete rotation about their common center of gravity, O.

The first question to be asked, under such circumstances, is whether it is possible for body A and B to move in any way so that the orbits forever remain circles of constant diameter.

I maintain that this is not possible. For suppose that it were possible to adjust the speeds of bodies A and B very slightly to accommodate the fact that the forces did not act though the common center of gravity in such a way as to still keep bodies A and B forever rotating in a circle of constant diameter about the immovable common center of gravity.

Now in order to maintain a body in an unchanging circular orbit, a force must be applied to the body in such a way that the component of the force in the direction from the body to the immovable center of gravity must exactly equal the force required to maintain the body in that circular orbit. This resultant force required must, as Newton showed, act in the direction from the body to the immovable center of gravity about which the body rotates.

But, in our example, the force of gravity acts at a small angle α from the line passing from the body to the immovable common center of gravity O. Suppose that we adjusted the speed of bodies A and B in their orbits so that the component of the force of gravity in the direction of the common center of gravity exactly equaled that required to keep bodies A and B in the circular orbit. Would this keep the bodies in an unchanging orbit? It would not. The reason for this is that even if the component of the force of gravity towards the center of rotation exactly equaled the force needed to keep the bodies in the circular orbit, due to the fact that the actual force of gravity acts at an angle α to the line between the body and the common center of gravity, there would therefore be a very tiny component of the force of gravity acting in a direction tangential to the orbit to consider. This tangential force can easily be computed from the laws of vector addition. If F is the total force of gravity between the two bodies, then the force tangential to the orbit would be $F.\sin(\alpha)$ or, in the case where alpha is very small, $F.\alpha$.

This tiny force tangential to the orbit, which might be too small to even be measurable directly, would none the less, by acting over a long period of time have a cumulative effect and make it's presence known by gradually increasing the orbital velocities of A and B. Once the orbital velocities of the two bodies increased, the force required to keep the bodies in the original circular orbit would increase. The gravitational force between the two bodies would then no longer be sufficient to maintain the bodies in a stable circular orbit. This would result in the two bodies moving apart slightly.

Hence a stable circular orbit is not possible.

In a similar way, if one considers elliptical orbits, one will find that whatever force acting towards the common center of gravity would be required, due to the fact that the forces of gravity do not act towards the common center of gravity, there would be a very small force acting in the direction tangential to line between the body and the common center of gravity which would over a long period of time tend to increase the speed of the bodies and hence make them move apart.

Hence stable elliptical orbits are not possible either.

And by a similar reasoning, one can show that whatever stable orbit one can conceive of, due to the fact that there would always be a small component of the gravitational force acting tangentially to the component of gravitational force required to keep the body in the stable orbit, the bodies would gradually speed up and therefore gain enough energy to move apart a measurable amount over a long enough period of time.

We can further extend our reasoning to the situation where the two bodies are not of equal masses and where the common center of gravity is not at the exact center of the line joining the two bodies. For in all cases of mutual action between the bodies, there will always be an additional component of the force tangential to the orbits of the bodies in excess to that required to maintain the bodies in stable orbits. These additional tangential forces will, therefore, alter the orbit of the body.

In all cases, therefore, if bodies are rotating about each other and being held in their orbits by actions at a distance that move at finite speeds, the bodies will slowly move apart.

Work and Energy considerations implied by bodies slowly moving apart due to the finite speed of gravity.

One of the hardest things that I had to come to terms with are the work and energy implications involved if bodies actually do move apart slowly as a result of the finite speed of gravity. In our everyday experience, large amounts of work have to be done to move a body from the surface of the Earth into deep space. If we had two bodies revolving about each other, as in the example shown above, classical mechanics enables us to work out how much work would be required to, for instance, increase the distance of the bodies from each other by a certain amount.

And yet, if gravity does move with a finite speed, all bodies rotating about each other will slowly drift apart. Where could this energy come from? We need to think outside of our existing formal training to answer this problem.

Perhaps a better way to look at the question of where does the energy come from would be to look at the problem in a different manner.

Suppose we have two bodies, such as the Earth and the Moon, orbiting about each other. Suppose that gravity did not exist and we were, instead, required to continually modify the direction of the Earth and Moon in such a way that they maintain the orbits that they currently do due to gravity. The amount of energy and work required to perform this feat is truly astonishing. And yet we have always assumed that gravity simply does all this work in some continuous and magical fashion.

Newton, in his Principia, discussed an example of whirling a stone about that was tied to a string. He did this to show that a force is continually required to keep a stone moving in a circular motion and that this force is provided by the string. We have, perhaps subconsciously, assumed since then that gravity behaves very much like a string in that it is capable of providing a steady pull on both bodies which depends upon the masses of the two bodies and the inverse of the square of the distance between them. In terms of analyzing problems concerned with gravity, gravity certainly does appear to behave in the same way that a string would.

The problem is that gravity is not a string. Although it may behave in some ways like a string, it obviously works in a completely different way altogether. My question is where does all the energy required to continually change the direction of the Earth and the Moon in their orbits about each other come from.

For in my analysis that follows, the energy involved when bodies slowly move apart is but a minute fraction of the energy required to keep bodies in their orbits. We have historically, in our analysis of gravity, used the term potential energy to describe the work, or energy, required to move a body against gravity or the work or energy that can be performed by a body in moving towards a source of gravity.

When a stone falls to Earth, we talk of potential energy being released which results in the stone gaining an equal amount of kinetic energy. We also know that when we lift a stone up, the amount of work that we have to do is the same as the amount of work that would be given out again if the stone fell to its original position.

Newton considered the pendulum and showed that the energy released when the pendulum moved toward the Earth was converted into kinetic energy and this energy was once again converted to potential energy when the pendulum reached the end of it's stroke. The fact that potential energy can be converted into kinetic energy and back again into potential energy has perhaps made it easy for us to simply not ask the more interesting question of what is the source of this potential energy due to gravity or where, for instance, is the potential energy gained by the pendulum stored. It is certainly not stored in the pendulum itself.

Consider a stone in deep space initially moving very slowly on a collision course with Earth. When it eventually strikes the Earth, it will have acquired a tremendous amount of kinetic energy. Two stones would acquire twice the energy. We are quite prepared to accept this, without question, as we know that this happens from practical experience. The stone from deep space striking Earth certainly did not acquire it's kinetic energy as a result of having originally been lifted from the Earth and moved to deep space.

The more fundamental question to ask, therefore, is when a stone falls where does this potential energy come from? I will try to answer this question in the next section where I present my theories on gravity and energy.

But for the moment, irrespective of whether or not the reader agrees with my theory of gravity, I would like the reader to consider that whatever the source of gravitational energy is, if it has the ability to perform work when a stone drops, then the same source of energy has the ability to cause two bodies, which should by our current theories never move apart, to actually move apart if the force of gravity does not act instantaneously but travels with a finite velocity.

And in the same way that we equate potential energy obtained from a stone falling with kinetic energy gained by the stone, so in the same way may we equate the kinetic energy gained due to the small component of the gravitation force moving tangentially to the orbit to the potential energy gained as the two bodies move further away from each other.

Estimation of the direction that bodies will move apart if gravity moves with a finite speed

Consider the case of two bodies rotating about each other in perfectly circular orbits. It is well known, from classical mechanics, that if both bodies were acted upon by a force of gravity moving at infinite speed that the bodies will rotate about each other about their common center of gravity.

Let us call the two bodies A and B and the masses of the two bodies M_A and M_B. Suppose the two bodies rotate about their common center of gravity O. Then if A and B are initially at positions A_1 and B_1 it is well known from classical mechanics that

$$M_A.\text{distance}(A_1 O) = M_B.\text{distance}(B_1 O)$$

Let us assume that in the time that it takes gravity to move from A and B and from B to A that the two bodies will have moved from points A_1 and B_1 to A_2 and B_2 as shown in Section 1 Figure 5 below:

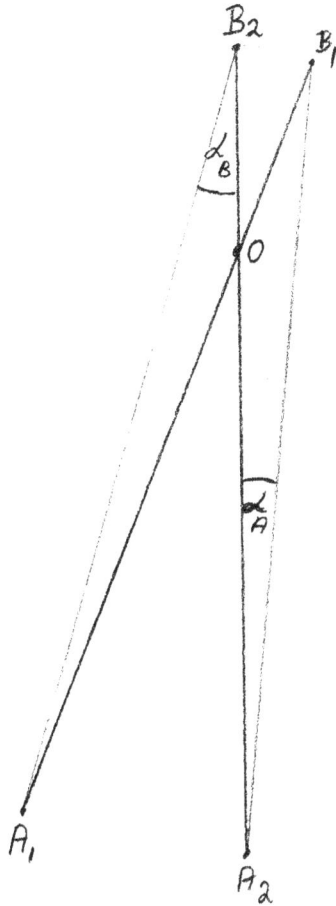

Section 1 - Figure 5

Let us further assume that these distances moved, A_1A_2 and B_1B_2 are extremely small in comparison to the distance A_1B_1, say less than a billionth of the distance. Then for all practical purposes, the error made in assuming that the common center of gravity, O, does not move during the time that it takes gravity to pass between the two bodies will be extremely small.

Now when body A arrives at point A_2, the force of gravity reaching A at this time will be that which resulted from B when B was at position B_1. Similarly, the force of gravity reaching B at point B_2 will be that which resulted from A when A was at point A_1.

$$\text{Let } \alpha_A = \angle B_1 A_2 B_2 \text{ and } \alpha_B = \angle A_1 B_2 A_2$$

Now if $A_1 A_2$ and $B_1 B_2$ are extremely small in comparison to the distance $A_1 B_1$ then we can assume, with a high degree of accuracy, that

$$A_1 B_2 = A_1 B_1 = A_2 B_2 = A_2 B_1$$

Also

$$\frac{A_1 A_2}{B_1 B_2} = \frac{A_1 O}{B_1 O} = \frac{M_B}{M_A}$$

since triangles $A_1 O A_2$ and $B_1 O B_2$ are similar, being almost perfect isosceles triangles with equal and opposite angles.

Now $\alpha_A = \angle B_1 A_2 B_2 = \dfrac{B_1 B_2}{A_2 B_2}$ in radians very nearly as α_A is very small.

Similarly $\alpha_B = \angle A_1 B_2 A_2 = \dfrac{A_1 A_2}{A_2 B_2}$ in radians very nearly as α_B is very small.

Hence

$$\frac{\alpha_B}{\alpha_A} = \frac{A_1 A_2}{B_1 B_2} = \frac{A_1 O}{B_1 O} = \frac{M_B}{M_A}$$

Now if angles α_A and α_B are very small and the bodies A and B are moving apart very slowly, angles α_A and α_B will change just as slowly and can, certainly for a few rotations of the bodies about each other, be treated as constants.

Hence, as body A moves about point O in an almost perfect circular orbit, angle α_A can be regarded as a constant angle between the line joining body A and B, passing through O and the direction of the force of gravity from body A acting upon body B at any instant of time. Hence angle α_A can be regarded as a constant angle by which the gravitational force of body B upon body A deviates from the direction in which the gravitational force would act if gravity moved with infinite speed.

Similarly, angle α_B can be regarded as a constant angle by which the direction of the gravitational force of body A upon body B deviates from the direction in which the gravitational force would act if gravity moved with infinite speed.

Obviously, over long periods of time the angles α_A and α_B will change depending upon the distance between A and B. The smaller angles α_A and α_B are, however, the less they will change in a couple of rotations. In our analyses, it is the values of angles α_A and α_B rather than the slight changes in angles α_A and α_B that are important as I will show shortly.

Let us now examine in more detail what would happen to body A as it moves in a circular orbit about the common center of gravity O once it arrives at point A_2, as shown in figure 6 below.

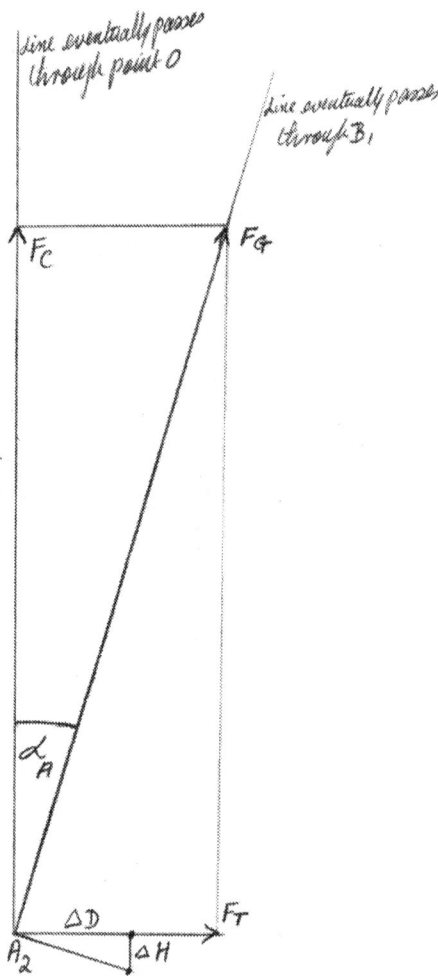

Section 1 - Figure 6

Suppose the force required to keep the body moving in a circular orbit about center O is F_C acting at any instant of time in the direction of the straight line from A through O. Suppose that A in the diagram is moving in a circular anti-clockwise direction. Suppose that this force, F_C, is supplied by gravity which moves at a finite speed. Then as explained above, the force of gravity would deviate by a certain small angle from the direction of force F_C. We called this angle of deviation α_A. Let the force of gravity acting upon A at point A_2 be F_G. Then force F_G acts in the direction of line $A_2 B_1$ and must be such that the component of force F_G acting in the direction from A_2 to O must equal force F_C which is required to keep body in a circular orbit about O.

Then there will also be a component of force F_G acting in a the direction of the tangent to the circle passing through point A_2. Let us call this tangential force F_T. Now by the laws of vector addition, $F_T = F_G . \sin(\alpha_A) = \alpha_A . F_G$ when α_A is very small. Also, when α_A is small, F_G is very nearly equal to F_C.

Hence when α_A is very small,

$$F_T = \alpha_A . F_C$$

or

$$\frac{F_T}{F_C} = \alpha_A$$

Now let us imagine that body A moves a small distance Δd around the circle from point A_2. Let this distance Δd be very much less than the distance $A_1 A_2$ which body A moves during the time that it takes gravity to pass from A to B. During the time taken to move the distance Δd the tangential force F_T will impart a certain amount of kinetic energy to body A causing it's speed to increase slightly. This kinetic energy gained is simply the force times the distance over which it was applied, or

$$\text{gain in kinetic energy} = F_T . \Delta d$$

This gain in kinetic energy is due to body A having increased it's velocity slightly due to the tangential component, F_T, of the force of gravity, F_G, acting on body A.

Now since body A has increased it's velocity, it will move away from O a certain amount. Suppose we assume that at the end of moving distance Δd around the circle, body A then moves away from point O a small distance Δh in the direction of the straight line from O through the position that A reached after distance Δd as shown in the diagram above. Suppose this slight movement Δh adjusts the speed of body A so that it's velocity around the circle returns to the original velocity that it had at point A_2.

What we are effectively saying is that we will convert all the increased kinetic energy gained during distance Δd into potential energy by moving body A away from O by a distance Δh.

Now the energy required to move A away from O by a distance Δh is, once again, simply the distance multiplied by the force being overcome. But the force being overcome is F_C. Hence the work done to increase the potential energy is

$$\text{Increase in potential energy} = F_C.\Delta h$$

But, if body A is to resume it's original circular velocity, the gain in kinetic energy is all used in increasing the potential energy. Hence

$$F_T.\Delta d = F_C.\Delta h$$

or

$$\frac{\Delta h}{\Delta d} = \frac{F_T}{F_C}$$

but as shown earlier,

$$\frac{F_T}{F_C} = \alpha_A$$

Hence

$$\frac{\Delta h}{\Delta d} = \alpha_A$$

But, looking at the diagram, as Δd is decreased, one can see that $\frac{\Delta h}{\Delta d}$ represents the angle between the tangent at point A_2 and the direction in which body A moves after a distance Δd. Hence body A will spiral away from the common center of gravity with the angle α_A. But α_A is also the angle between body A and the common center of gravity and body A and the actual direction of the force of gravity. Hence the direction in which body A would move at any instant of time is at right angles to the direction to the force of gravity, F_G, acting upon the body at that instant of time.

Now if α_A is very small, during one rotation of body A about the center of gravity O, angle α_A will be almost constant during the entire rotation. In addition, the distance from A to center O is almost constant. Hence, for one rotation of body A about the common center of gravity O, the total distance, H_A, that body A would move away from O would be

$$H_A = \alpha_A.2\pi.A_1O$$

Now consider, once again, the two bodies A and B with masses M_A and M_B rotating about their common center of gravity O discussed earlier.

Let H_A be the distance that body A moves away from the common center of gravity O during one complete rotation about O. Let H_B be the distance that body B moves away from the common center of gravity O during one complete rotation about O.

Then

$$H_A = 2\pi.A_1O.\alpha_A$$

and

$$H_B = 2\pi.B_1O.\alpha_B$$

But, as shown earlier,

$$\frac{\alpha_B}{\alpha_A} = \frac{A_1A_2}{B_1B_2} = \frac{A_1O}{B_1O} = \frac{M_B}{M_A}$$

Hence

$$B_1O.\alpha_B = A_1O.\alpha_A$$

Therefore

$$H_B = 2\pi.B_1O.\alpha_B = 2\pi.A_1O.\alpha_A = H_A$$

We thus arrive at the rather interesting conclusion that irrespective of the relative masses of the two bodies A and B, if they move apart from their common center of gravity O they will each instantaneously move apart by the same distance.

Hence if bodies A and B move apart from each other a total distance H_{AB} during one rotation about their common center of gravity, each body will have moved apart one half that distance.

Now

$$H_{AB} = H_A + H_B = 2H_A = 2H_B = 4\pi.A_1O.\alpha_A = 4\pi.B_1O.\alpha_B$$

and hence

$$\alpha_A = \frac{H_{AB}}{4\pi.A_1O} = \frac{\text{Distance moved apart in one revolution}}{\text{Twice orbital circumference of path of body } A}$$

and

$$\alpha_B = \frac{H_{AB}}{4\pi.B_1O} = \frac{\text{Distance moved apart in one revolution}}{\text{Twice orbital circumference of path of body } B}$$

Calculation of the Speed of Gravity.

We are fortunate in that of all the planets in our solar system, apart from Pluto, the Earth has the largest moon in relation to its size of all the planets. We are even more fortunate that there is only one moon orbiting the Earth. And most of all, we are extremely fortunate that a laser reflector was placed upon the Moon's surface by two of the Apollo 11 astronauts, Dr. Aldrin and Dr. Armstrong, in July 1969 enabling extremely accurate measurements to be made of the distance between the Earth and the Moon.

It is my understanding that measurements made over the last thirty-seven years indicate that the Earth and Moon are moving apart at a distance of about one and a quarter inches per year. I have read of various explanations for this phenomenon which attempt to show that this is as a result of action of tides etc. I do not agree with these ideas. For to move the Earth and Moon apart should, according to our current theories, require energy. To suggest that this energy can be obtained from tidal effects does not make sense to me. For any tidal effects should result in energy being lost as a result of friction. This is a net energy loss, not a gain. This energy loss should result in the speed of rotation of the Earth upon it's axis gradually decreasing until some time in the distant future the Earth and Moon would orbit each other with the same side of the Earth facing the Moon at all times. Or alternatively this energy loss should, if it were possible in some mysterious way to make up this energy loss by reducing the potential energy of the Moon with respect to the Earth, result in the Moon moving closer to the Earth, not further away. For if it were possible for any means to exist by which an energy loss due to tides could be converted into a gain in potential energy, we should be able to duplicate this mechanism and, by merely wasting energy, be able to rise off the Earth.

My feeling is that the Earth and Moon are moving away from each other due to the fact that gravity does not act instantaneously but rather travels with a finite speed. I have used this measured speed at which the Earth and Moon are moving apart as a method to calculate the speed of gravity. In the discussion which follows, I will use the most accurate values of measurements made that are available merely to prevent errors from arising in the calculation of the answer. The answer itself cannot, however, be more accurate than the least accurate measurement that I have available, namely the rather rough figure quoted of one and a quarter inches per year or 3.175 cms per year.

My purpose is also to get an approximation to the answer using my rather limited mathematical expertise. I will leave it to others, with better abilities than me, to take into consideration the effects caused by the fact that the Moon moves in an elliptical orbit around the Earth.

The first assumption that I will make in arriving at a first approximation is to assume that the Earth and the Moon revolve in circular orbits about a common center of gravity.

The second assumption that I will make, in view of the fact that a distance of one and a quarter inches in a year in a distance of 230,000 miles is a very small amount and that one can, therefore, use the formulae that I developed earlier in this section titled **"Estimation of the direction that bodies will move apart if gravity moves with a finite speed"**

For brevity in what follows, I will refer to the Earth and Moon rotating exactly once about each other, with respect to the background stars, as a rotation.

The sidereal month or time that it takes the Moon to orbit around the Earth once with respect to the background stars, is 27.321662 days or 2360591.5968 seconds.

There are 365.24 days in a year. Hence the Moon and Earth each rotate about their common center of gravity 365.24 / 27.321662 or 13.36814722 times in a year.

The Earth and Moon move apart 1.25 inches or 3.175 cms in a year.

Hence the Earth and Moon move apart 3.175 / 13.36814722
= 0.237504865 cms per rotation.
= 2.37504865E-6 Kms per rotation
This is about one eleventh of an inch per rotation.

The mass of the Earth is 5.9736E+24 Kgs.
The mass of the Moon is 7.349E+22 Kgs.
The ratio of the mass of the Earth to the Moon is thus 5.9736E+24 to 7.349E+22 or 81.28452851 to 1.

Hence the distance of the Moon from the common center of gravity of the Earth and Moon is 81.28452851 times the distance of the Earth from the common center of gravity.

The mean distance between the Earth and the Moon is 384400 Kms.

Hence the mean distance of the Moon from the common center of gravity is:
           81.28452851 / (81.28452851 + 1) * 384400 Kms
                     = 379728.4049 Kms.

The mean distance of the Earth from the common center of gravity is:
    1 / (81.28452851 + 1) * 384400 Kms = 4671.595098 Kms.

Assuming that the Earth and Moon move in circular orbits about the common center of gravity, the distance that the Moon travels around the common center of gravity per rotation is thus $2*\pi*$ 379728.4049 or 2385903.9344 Kms per rotation. This is the distance of the orbital circumference of the Moon about the common center of gravity.

The distance that the Earth moves around the common center of gravity per rotation is $2*\pi*$ 4671.595098 or 29352.49768 Kms per rotation. This is the distance of the orbital circumference of the Earth about the common center of gravity.

In the previous section, titled **"Estimation of the direction that bodies will move apart if gravity moves with a finite speed"**, the following formula was developed to give the angle of gravitational deviation for a body moving in a circular orbit:

$$\alpha_A = \frac{H_{AB}}{4\pi.A_1O} = \frac{\text{Distance moved apart in one revolution}}{\text{Twice orbital circumference of path of body } A}$$

So, considering the Moon,

$$\alpha_{moon} = \frac{2.37504865E-6}{2*2385903.9344} = 4.97725121E-13 \text{ radians}$$

Now considering the diagram below, this tangential angle α_{moon} is the same as the angle between the line joining the Earth and the Moon through the common center of gravity and the line joining the Moon to where the Earth was when gravity left it on its way to the Moon.

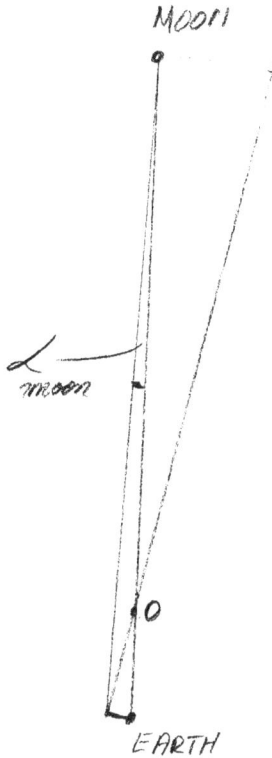

Section 1 - Figure 7

So the distance the Earth moved in the time that it takes gravity to move from the Earth to the Moon is α_{moon} * distance from Earth to Moon = α_{moon} * 384400 kms.

This is equal to 4.97725121E-13 * 384400 kms

$$= 1.913255E-07 \text{ Kms.}$$

As shown above, the Earth moves about the circumference of its orbit a distance of 29352.4976 Kms per rotation. One rotation takes 2360591.5968 seconds.

So the Earth moves at a speed of 29352.4976 / 2360591.5968 kms/sec or about 0.01243438 Kms/sec about the common center of gravity.

Now the Earth moves 1.913255E-07 Kms in the time that it takes gravity to move from the Earth to the Moon. Since the Earth is moving at a speed of 0.01243438 Kms/sec, the time taken for the Earth to move 1.913255E-07 Kms is 1.913255E-07 / 0.01243438 = 1.53868E-05 seconds.

Hence the time that it takes for gravity to move from the Earth to the Moon is 1.53868E-05 seconds.

Hence gravity moves the distance of 384400 kms in 1.53868E-05 seconds.

Hence the speed of gravity is approximately 384400 / 1.53868E-5 kms / second = 24,982,428,018 kms/sec.

This is approximately = 25,000,000,000 kms/sec or 2.5E+10 kms/sec.

The speed of light is 299,742.458 kms/sec.

Hence gravity moves at very roughly 83,000 times the speed of light.

This means that one light year is approximately 378 gravity seconds or about six and a quarter gravity minutes.

Our sun is about 149 million kilometers from us. Hence it takes gravity about 6 thousandths of a second to reach us from the sun.

Our closest star is about 4 light years away. So it takes gravity about 25 minutes to move from us to our closest star.

And it takes gravity about one year to move across our galaxy.

What is the maximum time that the Moon could have been moving away from the Earth?

The Moon has been measured to be moving away from the Earth at a rate of about 1.25 inches per year or about 20 miles in a million years. If we assumed that the Moon had been traveling away from the Earth at a constant rate, one would find that it would have taken about 12 billion years for the Earth and Moon to move apart about 240,000 miles.

The assumption that the Earth and Moon are moving apart at a constant rate is, as I will show, incorrect if the reason for the Earth and Moon moving apart is due to the speed of gravity. The purpose of this section is to determine that maximum time that the Moon could have been moving apart from the Earth if one assumes that the reason for moving apart is due to the finite speed of gravity.

The first thing that needs to be done is to determine how the speed at which the Earth and Moon move apart from each other would change as the distance between the Earth and Moon changes.

Once again, I will assume as a first approximation that the Earth and the Moon revolve about each other in almost perfect circular orbits.

Looking at the figure below, let us assume that at any instant of time the Moon is at point A, the Earth is at point B, and the common center of gravity is at O. As explained previously, let us assume that when the Moon, at point A, receives gravity from the Earth the gravity originally left the Earth when the Earth was at point C.

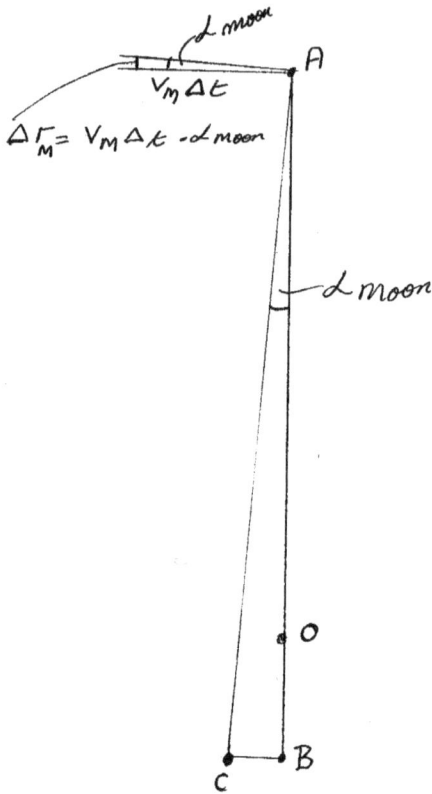

Section 1 - Figure 8.

Let M_E be the mass of the Earth.

Let M_M be the mass of the Moon.

Let V_G be the velocity of gravity.

Let V_E be the velocity of Earth in it's orbit about the common center of gravity O.

Let V_M be the velocity of Moon in it's orbit about the common center of gravity O.

Let $\alpha_{moon} = \angle CAB$, the angle of deviation between the direction of the Earth's gravity upon the Moon and the direction of the line between the Earth and the Moon.

Let r be the distance between the Earth and the Moon. In the figure above this is equal to the length of line AB.

Let I_g be the interval of time taken for gravity to travel the distance r from the Earth to the Moon.

Looking at the figure above:
$$AC = I_g.V_g$$
and
$$CB = I_g.V_E$$
Hence
$$\alpha_{moon} = \frac{CB}{AC} = \frac{I_g.V_E}{I_g.V_g} = \frac{V_E}{V_g}$$
In a similar way it can be shown that
$$\alpha_{earth} = \frac{V_M}{V_g}$$
As shown previously, the angles of deviation, α_{moon} and α_{earth} are also the angles at which the Moon and Earth deviate in their motions from the tangents to their circular orbits.

Let Δt be a small interval of time during which the distance between the Earth and the Moon increases a small distance Δr. So

$$\Delta r = \alpha_{moon}.V_M.\Delta t + \alpha_{earth}.V_E.\Delta t$$

Substituting in the values for α_{moon} and α_{earth} found earlier gives:

$$\Delta r = \frac{V_E}{V_g}.V_M.\Delta t + \frac{V_M}{V_g}.V_E.\Delta t = \frac{2.V_M.V_E}{V_g}.\Delta t \qquad (1)$$

Now if one considers the force acting upon the Earth and Moon it is very nearly given by Newton's formula as

$$F = \frac{G.M_E.M_M}{r^2}$$

Now this force is, as shown previously, very nearly equal to the force required to keep the Moon and the Earth in circular orbits about the common center of gravity.

Now, as is well known from classical mechanics, the force required to keep the Moon in its circular orbit about the common center of gravity is given by the formula:

$$F_M = M_M.\frac{V_M^2}{AO}$$

and the force required to keep the Earth in its circular orbit about the common center of gravity is:

$$F_E = M_E.\frac{V_E^2}{BO}$$

And as is also known, the above two forces must each be equal to the mutual force of gravity between the Earth and the Moon.

Hence

$$F = F_M = F_E$$

or

$$F = \frac{M_M.V_M^2}{AO} = \frac{M_E.V_E^2}{BO} = \frac{G.M_E.M_M}{AB^2} \qquad (2)$$

Now, due to the fact that O is the common center of gravity

$$M_M . AO = M_E . BO$$

or

$$AO = \frac{M_E}{M_M} . BO$$

Now

$$AO + BO = AB$$

so

$$\frac{M_E}{M_M} . BO + BO = AB$$

and hence

$$BO = \frac{M_M . AB}{M_M + M_E}$$

Now substituting the value of BO into the force equation (2) above yields

$$F = \frac{M_E . V_E^2 . (M_M + M_E)}{AB . M_M} = \frac{G.M_E.M_M}{AB^2}$$

Hence

$$V_E^2 = \frac{AB.M_M.G.M_E.M_M}{AB^2.M_E.(M_M + M_E)}$$

Now M_M, M_E and G are all constants in the above equation. Hence

$$V_E^2 = \frac{k2}{AB} = \frac{k2}{r}$$

or

$$V_E = \frac{k3}{\sqrt{r}}$$

where $k2$ and $k3$ are constants.

In a similar fashion it may be shown that

$$V_M = \frac{k4}{\sqrt{r}}$$

where $k4$ is a constant.

So the tangential velocities of the Earth and Moon moving about their common center of gravity are inversely proportional to the square root of the distance between the Earth and the Moon.

Now as was shown earlier in (1), the distance that the Earth and Moon move apart in a small interval of time is given by the equation

$$\Delta r = \frac{2.V_M.V_E}{V_g}.\Delta t$$

But, as has been shown, the velocities are inversely proportional to the square root of the distance between the Earth and Moon. Substituting the values for V_M and V_E into the above equation gives:

$$\Delta r = \frac{2.V_M.V_E}{V_g}.\Delta t = \frac{2}{V_g}.\frac{k3}{\sqrt{r}}.\frac{k4}{\sqrt{r}}.\Delta t = \frac{k5}{r}.\Delta t$$

where k5 is a constant. Rearranging yields:

$$\Delta t = K_g.r.\Delta r$$

where K_g is a constant.

Hence the time taken for the Earth and Moon to move apart a certain distance, due to the finite velocity of gravity, is proportional to the distance between the Earth and Moon. Hence, when the Moon and Earth were twice as close to each other as they are today, they moved apart at twice the rate.

When I first worked this out, I was surprised as I had intuitively expected that the further apart the bodies were, the longer it would take gravity to pass between them and hence the faster the bodies would move apart. This is not so, however, due to the fact that the orbital velocities are inversely proportional to the square root of the distance apart. So if, for instance, the distance between the bodies doubled, the time taken for gravity to pass between the bodies would double. But the orbital velocities would only be $\frac{1}{\sqrt{2}}$ of the original orbital velocities. So the distances covered by the bodies in their circular orbits during twice the time that it took gravity to pass between the bodies would only be $\frac{2}{\sqrt{2}} = \sqrt{2} = 1.414$ the distance covered originally in the time that it took gravity to pass between the bodies. The angle of deviation would thus be only .7071 the original angle of deviation. So if we double the distance between the bodies, the total circumference of the orbital path doubles. As the angle of deviation is only 0.7071 of the original angle of deviation, the total distance moved apart in one revolution would be two times 0.7071, i.e. $\frac{2}{\sqrt{2}} = \sqrt{2} = 1.414$ of the distance moved apart in one revolution at the original distance. So although intuitively one is correct in that the bodies will move apart further in one complete revolution as the distance between the bodies increases, because of the fact that the time for one revolution increases as the bodies move further apart, the bodies will actually move apart at a slower rate as the distance between them increases.

Let us now go back to our equation
$$\Delta t = K_g . r . \Delta r$$
where K_g is a constant.

We could work out the value of K_g very simply by substituting the measured value at which the Earth and Moon are currently moving apart.

Currently the Earth and Moon are moving apart at the rate of 1.25 inches per year, or about 3.175E-05 kms. per year. The Earth and Moon are currently about 384,400 kms apart. Substituting these values into the above equation yields:

$$K_g = \frac{\Delta t}{r.\Delta r} = \frac{1}{384400 * 3.175E - 05} = 0.0819356 \text{ years/km}^2$$

Some mathematicians might feel uneasy about me simply substituting values directly into the above equation. This is something that I do quite regularly in situations where our measured periods and distances are small compared to the total distances and times involved.

A more mathematically correct way would be to perform the integration and from that obtain the constant K_g that we are looking for. Starting with

$$\Delta t = K_g.r.\Delta r$$

gives the differential equation

$$dt = K_g.r.dr$$

Let us now integrate this equation for a period of one year from the present. Assume that the current distance between the Earth and the Moon is r kms and our current time is 0 years. Then in one year's time out new time will be 1 and the new distance from the Earth to the Moon will be $r + 3.175E - 05$ kms.

Integrating the above equation with these boundary conditions gives:

$$\int_0^1 dt = \int_r^{r+3.175E-05} K_g.r.dr$$

Hence

$$[t]_0^1 = K_g.\left[\frac{r^2}{2}\right]_r^{r+3.175E-05}$$

Hence

$$(1-0) = \frac{1}{2}.K_g.((r+3.175E-05)^2 - r^2)$$

or

$$1 = \frac{1}{2}.K_g.((r^2 + 2*r*3.175E-05 + (3.175E-05)^2) - r^2)$$

or

$$1 = \frac{1}{2}.K_g.(2*r*3.175E-05 + (3.175E-05)^2))$$

Hence

$$1 = K_g.(r*3.175E-05 + \frac{1}{2}*1.008E-09)$$

Hence

$$1 = K_g.(r*3.175E-05 + 5.04E-10)$$

Now r has a value of 384400. Substituting this into the above equation yields:

$$1 = K_g.(384400*3.175E-05 + 5.04E-10)$$

Hence

$$K_g = \frac{1}{12.2047 + 5.04E-10}$$

or

$$K_g = 0.0819356 \text{ years/km}^2$$

This is exactly the same result that was obtained earlier by simply substituting values into the equation

$$\Delta t = K_g.r.\Delta r$$

Lets now get back to our differential equation

$$dt = K_g.r.dr$$

where K_g has been determined to be 0.0819356 years/km^2

Lets now integrate this equation to determine the maximum time that the Earth and Moon could have been moving apart. Assume that some colossal impact gouged out a part of the original Earth splitting the Earth and Moon into two separate bodies. The closest apart that the Earth and Moon could have been after the impact would have been if the Earth and Moon were in perfect circular orbits about each other where their surfaces were almost touching. The diameter of the Earth is 12742 Kms. The diameter of the Moon is 3647 Kms. The closest that they could have been as separate bodies is thus to have their centers a distance of (12742+3647)/2 = 8194.5 kms apart. At this point their surfaces would have been just touching.

Suppose our current time, when the Moon is 384,400 Kms apart from the Earth, is time 0. Suppose at time $-Tmx$, the time immediately after the impact, the centers of the Earth and Moon were 8194.5 Kms. apart. Then during this maximum interval of time the Earth and Moon have been slowly drifting apart to their current distance of 384400 Kms. apart.

Integrating the above equation with the above boundary conditions gives:

$$\int_{-Tmx}^{0} dt = \int_{8194.5}^{384400} K_g . r . dr$$

Hence

$$[t]_{-Tmx}^{0} = K_g . \left[\frac{r^2}{2} \right]_{8194.5}^{384400}$$

or

$$0 - (-Tmx) = \frac{1}{2} . K_g . (384400^2 - 8194.5^2)$$

or

$$Tmx = \frac{1}{2} . 0.819356 * (384400^2 - 8194.5^2) \text{ years}$$

Hence
$$Tmx = 6.05E + 9 = 6.05 \text{ billion years}$$

This means that if the Earth and Moon were initially so close that their surfaces were nearly touching, they would have taken six billion years to have moved apart to their current distance from each other. Some assumptions that I made in arriving at the above figure are that the masses of the Earth and Moon have not changed significantly since the original impact that formed them, that the average speed of gravity has remained constant during the last six billion years and that the constant G, used in Newton's law of gravitation, has remained the same over the last six billion years.

If it can be shown in any way that the Moon and Earth have been orbiting about each other for more than six billion years then either my concepts of gravity are wrong or gravity moves even faster than I have supposed and there is some other reason for the Earth and Moon moving apart.

Tables showing how the Earth and Moon may have moved apart with time.

As shown above the equation
$$dt = K_g.r.dr$$
can easily be integrated and solved for any boundary conditions. Let R kms. be the distance that the Earth and Moon were apart at a time T years ago. Integrating this equation with these boundary conditions would give:

$$\int_{-T}^{0} dt = \int_{R}^{388400} K_g.r.dr$$

Hence

$$[t]_{-T}^{0} = K_g.\left[\frac{r^2}{2}\right]_{R}^{384400}$$

or

$$T = \frac{1}{2}.K_g.(384400^2 - R^2) = \frac{1}{2}*0.0819356*(384400^2 - R^2) \text{ years}$$

or

$$T = -0.0409678R^2 + 6.05353978*10^9$$

The following two tables, shown as Section 1 – Tables 2A and 2B, based upon the above equations, show the approximate distance from the Earth to the Moon at various periods in the past on the basis that the Earth and Moon are moving apart due to the finite speed of gravity.

Distance Apart (Kms)	Years Ago		Years Ago	Distance Apart (Kms)
8,200	6.05E+09		6.05E+09	9,295
10,000	6.05E+09		6.00E+09	36,151
50,000	5.95E+09		5.50E+09	116,239
100,000	5.64E+09		5.00E+09	160,363
150,000	5.13E+09		4.50E+09	194,733
200,000	4.41E+09		4.00E+09	223,888
250,000	3.49E+09		3.50E+09	249,661
260,000	3.28E+09		3.00E+09	273,011
270,000	3.07E+09		2.50E+09	294,516
280,000	2.84E+09		2.00E+09	314,554
290,000	2.61E+09		1.50E+09	333,391
300,000	2.37E+09		1.00E+09	351,218
310,000	2.12E+09		5.00E+08	368,183
320,000	1.86E+09		0.00E+00	384,400
330,000	1.59E+09		-5.00E+08	399,960
340,000	1.32E+09		-1.00E+09	414,937
350,000	1.03E+09		-1.50E+09	429,392
360,000	7.44E+08		-2.00E+09	443,376
370,000	4.45E+08		-2.50E+09	456,932
380,000	1.38E+08		-3.00E+09	470,097
384,400	0.00E+00		-3.50E+09	482,904

Section 1 – Table 2A   **Section 1 – Table 2B**

It can be seen from Table 2A on the left side table that the Earth and Moon would have moved apart very rapidly during the time period 6.05 to 5.6 billion years ago, moving from 8,200 kms apart to 100,000 kms apart in about 400 million years. It then took nearly a further 800 million years to move apart the next 100,000 kms to get to 200,000 kms apart. It then took over 1.9 billion years to move the next 100,000 kms to get to 300,000 kms. apart. And to move the final 84,400 kms apart then took a further 2.37 billion years.

Haley and Newton accurately predicted when Haley's comet would return, some decades after their deaths. If one could look at Table 2B above, the bottom seven lines estimate how far apart the Earth and Moon will be in the future. I would like to predict that in half a billion years from now, the Earth and the Moon will be about 400,000 kilometers apart and in three billion years from now they will be 470,000 kilometers apart. If this does indeed occur, then it will to some extent verify my ideas about gravity.

Some further thoughts on the Earth and Moon.

If one considers the size of the body required to split the Earth and Moon into two pieces, one will realize that the smaller the distance that the Moon and Earth were originally separated, the smaller the size of the body needed to collide with the original Earth to form the Moon. My feeling, on looking at the first of the two tables, is that the Earth and Moon were probably broken into two between 6.0 and 5.5 billion years ago.

As mentioned earlier, the equation used to derive the two tables above was based on the assumption that the Earth and Moon move in almost perfect circular orbits about each other.

Imagine that originally the Earth was slightly larger than it is today and that it had no moon. Now imagine that a large body, perhaps 1,200 miles to 1,800 miles in diameter, struck the original Earth a glancing blow and gouged out a portion of the Earth which then combined with some of it to form the Moon. It is very likely that the original orbit of the Moon about the Earth would have been highly elliptical with the Earth and Moon perhaps being no more than ten thousand kms apart at their closest part of the elliptical orbit. As shown earlier, the rate at which the Earth and Moon move apart is inversely proportional to their distance apart. If the Earth and Moon were in a highly elliptical orbit just after the collision that formed them, then during those times in the orbit when the Earth and Moon were closest together, the angles of gravitational deviation would have been larger and the Earth and Moon would have moved apart very much faster than during those times when the Earth and Moon were furthest apart in their orbit. This would have the effect that the Earth and Moon would move apart at a faster rate when they were closest to each other than they would when they were furthest away from each other. This would mean that not only would the Earth and Moon slowly move apart, but their original highly elliptical orbit would gradually move closer to a perfect circle, with the closest distance between the Earth and Moon increasing at a faster rate than the furthest distance.

If this is so, then the maximum time that the Moon has been moving away from the Earth will be less than the 6.05 billion years calculated earlier on the assumption that the Earth and the Moon were moving in circular orbits. I will leave it to others with more mathematical ability than I have to work out the mathematics to describe this. It should be possible, by considering the current elliptical orbit of the Earth and Moon, by working backwards using the concepts that I have illustrated, to determine what the original elliptical orbit of the Earth and Moon could have been to produce the current elliptical orbit of the Moon about the Earth. Once this is done, it should then be possible to get a good idea of what size of body must have struck the Earth and at what angle to produce the original elliptical path.

It is quite possible that this colossal impact which formed the Moon from part of the Earth also had the effect of tilting the axis of the Earth at an angle of 23 degrees to its plane of rotation about the sun. It might also be possible, once one can work out roughly what size body must have struck the original Earth, to work out a range of possible latitudes that this could have taken place that would have caused the axis of rotation to tilt by 23 degrees. It seems to me probable that the axis of the original Earth, before the collision which formed the Moon, would have been perpendicular to the plane of orbit of the original Earth about the sun. As a result of the collision and loss of matter in one portion of the spinning Earth, the axis of rotation might have changed to the current 23 degrees to the plane of orbit.

It may also be that the colossal impact which formed the Moon also had the effect of stirring up the core of the Earth in a way that might be responsible for the powerful magnetic field of the Earth.

Try to imagine, by looking again at the tables in the previous section, what conditions must have been like on the Earth when the Moon was very much closer to the Earth than it is today. Try to imagine what tides must have been like when the Earth and Moon were only a few tens of thousands of kilometers apart, particularly if their orbits were highly elliptical and the lunar month was only a few hours or days long. If there had been any atmosphere present in those days, the rapid approach and retreat of the Moon in a highly elliptical orbit would have stirred up the atmosphere considerably. It is quite possible that this would have caused far more intense electrical storms that we are accustomed to today, particularly if the Earth's magnetic field after the impact was much stronger that it is today. The light reflected from the Moon would have been much more intense than today. Eclipses of either the sun or Moon would have been very frequent, probably every couple of hours or days, on certain parts of the Earth.

But by 4.5 billion years ago, the Moon would have already moved to about 200,000 kms from Earth and its orbit would have been much more circular. Conditions then would, apart from considerably higher tides than today, a lunar month of about twenty days and much brighter nights during a full Moon, been more like what we are accustomed to today.

It may be that the body which collided with the original Earth to form the Moon contained a great deal of ice. If so, much of this ice might have vaporized on impact and been trapped by the Earth, eventually forming the oceans. If this did in fact happen, then this would explain a few questions that I have in my mind concerning the total amount of water present on the Earth. I have long suspected that the Earth contains far more water than one would expect for a planet so close to the Sun. If one looks at Mercury, Venus, Mars or our own Moon one will find that none of them have anywhere near the amount of water that the Earth has. If, as I suspect, all planets gradually move away from the sun over very long periods of time, then my feeling is that most of the water on the original Earth would have been burned off by the sun when the original Earth was much closer to the sun than it is today. In that case, the surface of the original Earth before the impact which created the Moon might have been much more like that of Mars with just a trace amount of water present. If the body which struck the original Earth was a ball of solid ice at least 1566 miles in diameter then it would have contained enough water to cover the entire Earth to a depth of about ten miles with water. It may be that during the last five to six billion years, much of this water has been burned off the Earth and the Moon by the sun. This would have gradually exposed land over the course of the last five to six billion years. The Earth might still be loosing water and might one day once again end up like Mars, Mercury or our own Moon.

The Earth might, in fact, still be in an unstable state as a result of the collision some six billion years ago and life might still be adapting to new and drier conditions. If this is so, the salinity in the seas will gradually increase over the next few billion years as the seas themselves gradually dry up. The fish and other life in the seas might continually be evolving to deal with higher and higher levels of salinity. It is quite possible that the very earliest life forms could not even survive in the seas of today.

My personal feeling is that when trying to imagine the conditions under which life initially evolved, one must never forget to take into account what highly unusual conditions were present on Earth shortly after the Earth was struck by the large body which produced the Moon. If these conditions had, as I suspect, any part in conditions favorable for the evolution of life, then life as we know it may be far less likely than previously imagined. In fact, the original conditions necessary for life to form may no longer be even be present on the Earth today and might have only been present for less than a hundred million years after the original Earth was struck by the large body which produced the Moon.

It may also be that the enormous impact which formed the Moon had the effect of bringing iron and other heavy elements up from the core of the Earth to the surface, making iron a fairly common element on the surface of the Earth and hence allowing life to evolve. For water is not the only thing required for life to evolve. Heavy elements such as iron are also required.

Thoughts on the perihelion of Mercury

The perihelion of Mercury has been measured to advance by about a second of an arc per century. The orbit of Mercury is more elliptical than any other planet except Pluto. In addition, Mercury is closer to the sun than any other planet and should, therefore, move away from the sun faster than any other planet. It may be that the perihelion is advancing due to the fact that gravity moves with a finite speed and Mercury is slowly moving away from the sun due to the small tangential component of the force of gravity acting upon Mercury. I have never investigated this, however, as the mathematics of elliptical orbits and how they would change as a result of the finite speed of gravity are a little beyond my abilities. I would appreciate it if those with better mathematical abilities than I have could investigate whether the perihelion of Mercury could be advancing as a result of the finite speed of gravity.

Conclusion

It has been shown in this section that if gravity moves with a finite speed then bodies in orbit about each other will slowly drift apart.

The Earth and the Moon have been measured to be slowly moving apart at a rate of about one and a quarter inches per year.

If the cause of this movement apart is, indeed, due to the fact that gravity moves with a finite speed, then it has been shown that the speed of gravity is about 2.5E+10 kms/sec or about 83,000 times faster than the speed of light.

And it has further been shown that if the cause of the Earth and Moon moving apart is due to the finite speed of gravity, then the maximum time that the Earth and Moon could have been moving apart is roughly six billion years. I would appreciate it if mathematicians with greater abilities than me could improve upon my calculations by properly taking into account the elliptical orbit of the Moon.

If the speed of gravity were limited to that of the speed of light, it would take gravity 83,000 times longer than the calculated value above to move between the Earth and the Moon. The Earth and Moon would, therefore, move about 83,000 times further around their orbits during the time that it took gravity to pass between the Earth and the Moon. Hence the angles of deviations would be 83,000 times bigger and hence the Earth and Moon would move apart about 83,000 times faster than calculated. So, if gravity moved at the speed of light, then instead of moving apart at about 1.25 inches per year, the Earth and Moon would move apart 83,000 times further, or about 1.58 miles per year. We would clearly be able to measure this distance. And in addition, the maximum time that the Earth and Moon could then have been moving apart would be 83,000 times less than the 6 billion years calculated earlier. This would be roughly 72,000 years. This is clearly inconsistent with everything that we have learned about our own evolution from fossil records.

So we are left with the alternatives that either gravity moves at a speed of about 83,000 times faster than light or that it moves even faster than that. Gravity certainly cannot move at the speed of light.

So, whatever the cause of gravity may be, gravity moves at least 83,000 times faster than light. And if gravity is a result of some physical process between bodies, then this process moves at least 83,000 times faster than the propagation of light. And if the physical process causing gravity in any way involves the transfer of matter or energy, then whatever form this transfer takes, something must be moving at least 83,000 times faster than the speed of light.

And anyone who claims that nothing can move faster than the speed of light must also assume that gravity in some way moves with infinite velocity, which is in itself a contradiction of the very claim that nothing can move faster than the speed of light. For whatever gravity is, we can rest assured that it works on physical principles. For to deny this is to merely admit that we do not understand the physical principles involved in the process called gravity.

This completes the purpose of this first section.

Section 2 - Theory on energy, matter and gravitation.

Propositions.

Proposition 1.
    Energy is matter in motion.

    All energy ultimately is matter in motion. Force is the result
    of matter in motion colliding with other matter and changing
    the momentum of all pieces of matter involved in the
    collision.

Proposition 2.
    Any transfer of energy is ultimately as a result of transfer of
    matter in motion.

Proposition 3.
    The aether is composed of a vast number of extremely small,
    small particles of matter moving in all directions at velocities
    significantly greater than the velocity of light.

    As I showed in Section 1, the velocity of gravity is some
    83,000 times faster than light. It is possible that some
    particles of matter in the aether travel at much higher
    velocities than this. In fact I see no reason why there should
    be an upper limit to the speed at which matter can move. My
    personal feeling is that the only "speed limit" which matter is
    subject to is due to the fact that the faster a piece matter
    moves, the sooner it will be before it collides with some other
    piece of matter which could break it up or slow it down.

Proposition 4.
    Gravity is not a property of matter itself. Gravity is instead a
    result of particles of matter in the aether colliding with very

much larger bodies of matter, moving at very much slower speeds, thereby imparting momentum to those bodies as they are reflected away after impact by those bodies.

Two bodies appear to be attracted towards each other by gravity due to the fact that each body is shielded by the other body from impacts of particles of matter in the aether that would otherwise occur if the other body was not present. This shielding effect results in each body being pushed by the aether towards the other body.

Proposition 5.

I postulate that matter has the following properties:
1. Matter is something which can exist in the absence of any energy and does not depend upon energy for it's existence.
2. Any piece of pure matter occupies a certain volume of space and cannot be compressed into a smaller volume of space.
3. No two pieces of matter can occupy the same volume of space at the same time.
4. A piece of pure matter is composed of nothing else than that pure matter. Any piece of pure matter can be split up into two or more smaller pieces of pure matter.
5. Pure matter is sub divisible in any proportions and there is no minimum size to which a piece of pure matter can be.
6. All matter has the property of inertia as explained by Newton.
7. Inertia is strictly cumulative for all matter. That is twice the amount of matter will have exactly twice the amount of inertia.
8. A piece of matter will move in its state of rest or uniform motion in a straight line, as defined by Newton, until it collides with another piece of matter. At that time both pieces of matter may combine, or break up and spread out in various directions subject to the laws of conservation of momentum as described by Newton and also subject to the laws of conservation of energy.

9. A piece of pure matter cannot absorb energy in any way other than by changing it's velocity.
10. Because of the fact that matter can exist in the absence of energy, pure matter in itself possesses no properties which bind it together. One piece of pure matter is bound together simply because of the fact that all of it's component portions are contiguous and travel together in the same direction and with the same velocity.
11. A piece of matter does not have any properties which in any way enable it to interact with any other piece of matter other than by colliding with it.

I visualize pure matter as being most closely analogous to a fluid with zero viscosity and zero surface tension. Try to imagine something more dense than anything that we have ever experienced and yet so soft that the slightest force would deform it. Try to visualize the smallest pieces of matter as things which cannot be compressed and cannot absorb energy other than by changing their velocities.

Proposition 6.

All the order and laws of physics which we are aware of are as a result of the random movement of matter at incredible speeds. Out of this chaos the order that we know grows as a result of physical principles and the laws of probability.

Proposition 7.

All forces commonly referred to as "Actions at a distance", such as gravity, magnetism and electrical attraction and repulsion, are not properties of matter itself but are rather a result of physical processes taking place in the aether. These processes do not act instantaneously but require a finite time to be transmitted between one body and another through the aether.

Introduction

Any theory that attempts to explain gravity must, at the least, not only explain why Newton's equation for gravity works but must also explain the three most important fundamental principles of gravity that Newton discussed in his General Scholium at the end of his Principia.

Firstly, how is it possible for the gravity from each little piece of matter from the farthest side of the sun to pass through something as large as the sun with no apparent resistance from all the other pieces of matter in the sun. For in all of our equations that have been derived for the centers of gravity of bodies, we assume that we can sum the effect of each little piece of matter in a body upon another body without having to take into account the effects of any intervening matter between the two bodies. These equations would not work as well as they do if the assumption was incorrect.

Secondly, why does the force of gravity depend upon the total mass of each body rather than the cross sectional area projected by each body, i.e. upon the area of the bodies projected upon each other. For if, as I suggest, gravity is as a result of a wind like effect of particles in the aether, the effect exerted upon a body should be proportional to the cross sectional area of the body rather than the mass of the body.

Thirdly, how is it possible that the aether offers no resistance to the planets in their orbits about the sun that would cause them to slow down. For, as Newton pointed out, the orbits of planets moving around the sun are virtually constant over very long periods of time. This could not happen if there was any significant resistance between the aether and the planets.

I will attempt to explain all of the valid questions which Newton raised. I would also like to offer my explanation as simply as possible so that someone without any scientific training can follow most of my explanation and at least get a general idea of the basic principles of my theory.

Thoughts on the aether.

Both Maxwell and Faraday, as well as many others going back to Plato's time and probably earlier, have come to the conclusion that an aether exists. Neither Maxwell nor Faraday was able to give a satisfactory explanation as to what the aether could be.

Newton, in his General Scholium at the end of his Principia, was keenly aware of the fact that he was unable to give an explanation for what gravity could be. Newton did make the profound observation that the force of gravity depended upon the total quantity of matter in each body and not upon the cross sectional area, or more accurately, the projection of each body upon the other. Newton also observed that this force of gravity could penetrate through even large bodies such as the sun.

Newton, in his day, could not imagine that bodies which we think of as solid are in fact far from solid. We now know that bodies consist of atoms which in turn consist of very dense nuclei and large amounts of virtually empty space between the nuclei. We are gradually learning that even protons, electrons etc. are made up of much smaller particles.

I propose that as one keeps on getting to smaller and smaller particles one will eventually arrive at particles of pure matter which are composed of nothing else than this pure matter, which I have discussed in Proposition 5. The density of this pure matter is something that is very hard to visualize.

I propose that the aether, which Maxwell and Faraday both accepted as existing in a form that they could not imagine, consists of nothing other than vast amounts of extremely small, small particles of pure matter moving in all directions at incredible speeds, of the order of magnitudes larger than the speed of light, through empty space. As was shown in the previous section, the speed of gravity is at least 83,000 times the speed of light. I do not know if gravity is a result of particles of matter directly colliding with other larger bodies or of some type of wave front which moves at 83,000 times the speed of light but which in itself is the result of much smaller particles of matter moving at even higher speeds. In view of the confusion that was caused due to Newton's corpuscular theory of light, we need to keep our minds open to both the possibility that gravity works in the same way as light does, i.e. by means of some wave action in the aether, or that it is a direct result of the collision of particles.

Let us, for example, visualize a given volume of one cubic centimeter of what we consider to be empty space filled with aether. At any instant of time, the total volume of matter, comprising the aether, in that cubic centimeter of space is incredibly small, perhaps no more than the volume of one or two atoms. If particles in the aether do move at 83,000 times the speed of light, they move at a speed of 2.5E+15 cms/sec. Hence any particle of the aether that does move through that cubic centimeter of space will only take at most four tenths of one thousand million millionth of a second to pass through our cubic centimeter of space. This means that there could be over two thousand five hundred million million particles of the aether moving through our cubic centimeter of space in a second without there ever being more than one of those two thousand five hundred million million particles of the aether in our cubic centimeter of space at any one time. When one starts thinking about this, one can understand that although the aether could be an almost perfect vacuum, the total mass of matter moving through a cubic centimeter of space in a second could be considerable, and, because of the enormous speeds at which the matter comprising the aether moves, the total energy that moves through that cubic centimeter of space in a second, in terms of what we understand as kinetic energy, could be so large as to be almost unimaginable.

I visualize the universe consisting of matter moving through empty space in all directions, colliding with other pieces of matter. Most of that matter in the universe may well be concentrated in large bodies such as galaxies, suns and planets which move relatively slowly. A small amount of matter, that which comprises the aether, moves at enormous speeds. Although the amount of matter is small, because of the enormous speed at which this matter moves as well as the enormous volume of empty space compared to the space occupied by galaxies of stars and planets, the kinetic energy carried by this matter is enormous and is spread throughout the aether and probably constitutes the repository for most of the energy in the universe.

This aether provides the mechanism by which all other larger scale phenomena, such as gravity and the transmission of light, occur. Newton in his time felt that light consisted of corpuscles moving through the vacuum of space. I disagree. I feel that light is released in discrete amounts, or quanta, as a result of forces acting on relatively large bodies such as protons, electrons, atoms or molecules. This energy, once released by an atom or molecule, is then transmitted away from the atom or molecule uniformly in all directions by the aether in much the same way as the energy from an explosion is carried away as a shock wave by our atmosphere. The aether acts as a medium enabling light to be transmitted in an analogous way to which our atmosphere allows sound to be transmitted.

The matter comprising the aether moves through all space in a completely random manner. However, over a period of time, the amount of matter moving in any particular direction would be equal to the amount of matter moving in any other direction. As an analogy, imagine a gas under pressure. Molecules of the gas move randomly but exert an equal pressure in all directions. If the gas did not exert an equal pressure in all directions, matter in the gas would be moved in the direction where it was not equal until such time as the pressure in all directions equalized.

In the same way, if the amount matter in the aether moving in any direction were less than in any other direction, matter moving and colliding in the aether would rearrange itself until the aether was back in equilibrium.

Thoughts on the possible amount of energy present in the aether.

I initially found it very hard to imagine that what we think of as empty space, with apparently nothing in it, can be the place where vast amounts of energy are stored. If you are in a room, look up and across to the far wall in your room. If you look at what is between you and the far wall, there does not appear to be anything between you and the wall. Now we know that there are a few pounds of air present but apart from that, the space appears to be empty. But it is not.

Imagine a machine gun firing a stream of ten lead bullets per second at a target. Suppose that the bullets are traveling at 1000 feet per second. Each bullet would be 100 feet apart from the next. Now most people cannot see bullets traveling at as little as 1000 feet per second. So when a machine gun is firing bullets, most people would not be able to see the bullets traveling even if they might hear the sounds of them whistling around their ears. Now imagine if you were to take each bullet and split it up into a hundred million small pieces and fire them so that the same amount of lead is fired per second. So instead of firing the bullets at a rate of a ten per second you were to fire the small pieces at a rate of a thousand million per second but at the same speed of 1000 feet per second. The total mass of lead and the kinetic energy being transferred would be the same as before. Now try to imagine trying to see these very small pieces leaving the machine gun at the rate of one thousand, million little pieces per second. There would be one million little pieces per foot. Even if the pieces were stationary, the pieces would each be so small that most people would be unable to see them. If they were traveling at one thousand feet per second then nobody would have a chance of seeing the pieces.

Now imagine that instead of firing the little pieces at a speed of one thousand feet per second, we could somehow fire them at a speed of eighty two million, million feet per second, the minimum speed that gravity travels through in the aether. The total mass transferred in a second would still be the same as before, namely the mass of ten lead bullets per second. But the total energy transferred would now be six thousand, seven hundred, million, million, million times as much as before or approximately that of sixty seven thousand million, million, million bullets per second. If we assumed conservatively that sixty seven thousand bullets contained the explosive power of a kilogram of TNT, the energy being transferred per second would be equivalent to a billion megaton nuclear bombs detonating every second. Furthermore, by increasing the velocity by a factor of eighty two thousand million times, each of the thousand million small pieces would be eighty two thousand million times further apart than before. The thousand million small pieces would now be spread over the distance of fifteen and a half thousand million miles which gravity covers in a second. This means that the pieces would now be separated by a distance of some fifteen and a half miles from each other.

In our example so far, the density of the little pieces of lead is still the same as it was originally. Let us now compress the lead so that it's density approaches that of pure matter. Each of these little pieces, invisible to the eye, would be more that fifteen miles apart. The volume of space occupied by these pieces in comparison to the total volume of space that they travel through in a second would be virtually negligible and would certainly form a better vacuum than anything that we could produce on Earth. And if we were then to further increase the number of pieces of matter and reduce their mass accordingly, so that each piece became much smaller in mass than the mass of a quark and so that the total mass always remained the same as that of ten bullets per second, each of the little pieces of matter would become so small that it could pass through the Earth with a high chance of not hitting anything at all. If we were to now fire this stream of matter through your big toe, the chance of any piece hitting any part of your toe would be virtually non existent. You would not even be aware of all of this energy passing through you.

Now imagine that instead of just one little stream of matter passing through your big toe, there were literally millions of streams of matter coming at you and passing through you equally from all directions through every little piece of your body. And through everything that you can see about you in all directions. This is what I visualize the aether to be.

If you are in a room look at the far wall again. The space between you and the far wall looks empty. Now try to imagine that there could be more energy passing through your room in all directions each second than that being released by the sun itself in a second. Now try to imagine that this energy, or matter in motion, passing through your room is so nearly in equilibrium in all directions that the only indication that we have that all this energy is present is that when you jump into the air, you keep on being pushed back to the floor very gently due to the slight imbalance in the equilibrium caused by the Earth itself shielding a very slight amount of the aether that would have otherwise have reached you by passing through the Earth to you.

Thoughts on what happens to pure matter during impacts.

Imagine that two pieces of pure matter collide with each other. What rules govern what happens to the matter after the impact?

Newton determined the rules which apply when regular objects collide. The rule that Newton determined is that when two bodies collide then the total momentum of all the matter before impact is the same as the total momentum of all the resulting pieces of matter after impact. The total momentum is obtained by multiplying the mass of each body by it's velocity, remembering that velocity is a vector quantity. As is well known, the law of conservation of momentum is, in itself, insufficient in determining the final velocities of each body. One could choose any particular final velocity of one of the bodies and, by using the law of conservation of momentum, find a final velocity for the other body that would satisfy the law of conservation of momentum.

The other thing that determines the final velocities of the bodies is the coefficient of restitution. If two equally large steel balls hit each other, the coefficient of restitution is close to 1 and the two balls will essentially bounce off each other, each transferring it's momentum to the other. If two equally large balls of something such as putty hit each other, the coefficient of restitution is very low and the two balls of putty might essentially stop each other in their tracks and essentially combine with each other to form one mass moving at zero velocity. If this were to happen, all the kinetic energy that the balls of putty had would be converted into heat and the balls of putty would increase in temperature.

The important thing to understand, in all types of collisions taking place between pieces of matter that we are familiar with is that there are in fact two laws of conversation that have to be obeyed. The first is the conservation of momentum and the second is the conservation of energy. Momentum is a vector quantity whereas energy is a scalar quantity. This is all well known and there are innumerable reference books that will explain this better than I can.

I would now like to ask the reader to open up his or her mind to the possibility that things that we take for granted might not always be quite as simple as they appear. We are all quite comfortable is accepting that kinetic energy can be converted into heat during an impact between two bodies. We know from experiments that the kinetic energy lost by the bodies during an impact will be equal to the heat energy produced. The tricky part comes when one asks what is this heat energy.

As I mentioned in Proposition 1, my feeling is that all energy is simply matter in motion, i.e. all energy is ultimately kinetic energy.

As I mentioned in Proposition 5, my feeling is that matter can exist in the absence of energy and that the only way that pure matter can "absorb" energy is by changing it's velocity. The concept of hot and cold cannot exist in pure matter as the only property that it has, other than occupying a volume, is it's inertia. What happens at the lowest level in a collision where pure matter cannot absorb energy in the form of heat?

My feeling is that at this lowest level the two laws of conservation, namely conservation of momentum as well as conservation of energy, must still both be obeyed. If one thinks about the implications of this in collisions of pure matter, where the only way that energy can exist is as kinetic energy, one will realize that collisions between pure matter will work in a way that is quite unlike anything that we are familiar with.

The first thing that will be totally different in collisions between pure matter and what we think of as normal matter is that the concept of a coefficient of restitution does not apply when considering impacts between pieces of pure matter. If one thinks about our everyday experiences with coefficients of restitution between normal matter, the coefficient of restitution in a way gives a measure of the ability of two bodies to absorb kinetic energy during an impact. The higher the coefficient of restitution, the less is the ability of the two bodies to absorb kinetic energy during an impact. And the lower the coefficient of restitution, the more is the ability of two bodies to absorb kinetic energy during an impact and to convert it into heat.

If one considers pieces of pure matter colliding, according to my postulates these pieces of pure matter have no ability to absorb any energy whatsoever other than by changing their velocities. This means, therefore, that the coefficient of restitution in any collision between two pieces of pure matter must be 1.

There is one other factor that must be taken into consideration when considering the impacts of pieces of pure matter. In all of our experiences with everyday compounds colliding, one of the most common things that we take for granted is that the bodies that we are familiar with require a certain force to break them up or to deform them. A steel ball is extremely difficult to break up into smaller pieces and even something as fragile as a drop of water still has surface tension, chemical bonds and gravity holding the component parts together.

When two pieces of pure matter collide, there is nothing to hold the component pieces of matter together at all. This is totally different to our everyday experience and is something that needs to be thought about to try to imagine what will happen in such collisions. The best way would be to perform experiments in space, with liquids with very low viscosity and surface tension, to try to get some feel and idea of what is likely to happen in collisions between particles of pure matter.

Thoughts on the times involved for processes to happen.

To most of us a second seems like a fairly short interval of time. However, those among us who have programmed computers realize that a tremendous number of computer instructions can be processed in a second. Think of a simple thing like a tune generated by a computer chip, such as one that one might find in a child's toy. Most people, merely listening to the sound generated by the toy, have no idea of how the sound was generated or of the millions of computer instructions that might have been executed to produce a sound lasting for a few seconds.

Indeed, if the most accomplished scientists from the nineteenth century were to be given the tools of their day to investigate the sound, it is doubtful if any of them could even begin to form an accurate guess as to how the sound in the child's toy was generated. The biggest hurdle to those scientists of a hundred years ago might well be their difficulty of understanding how a very complicated process could work so quickly to produce what appears to be a fairly simple result in so short a time.

Their real difficulty would be in understanding that even though a sound wave might vibrate a thousand times a second, the computer in the toy might be processing instructions at one thousand million instructions per second and that the computer could execute as many as a million instructions for each of the thousand vibrations of sound in the second.

Now think about light which vibrates at millions of billions of times per second. In my youth I felt that Newton's corpuscular theory must be correct. My biggest mistake at that time was that I tended to think that the propagation of light must be a fairly simple process. I now feel as a scientist of the nineteenth century, after having struggled in vain to find the reason for the sound out of the toy, might then feel after being given a clue that it is quite possible that some process could be involved such that for each vibration of a wave of sound, literally millions of other very much faster processes could have been completed which resulted in a single vibration of the source of the sound.

The question that I now ask myself when thinking about light is, is it possible that some other processes could be involved such that, for each vibration of a wave of light, millions of these other processes could have taken place and that cumulatively they could have resulted in just one vibration of a wave of light.

Try to imagine the kind of processes that might be involved from the time that a tiny piece of matter speeding through the aether strikes a piece of matter which is part of a great and slow moving body of matter such as an electron. If this is indeed the cause of gravity, then what we do know is that in some way this momentum gained from the collision must be transferred to all the matter in the electron and eventually to all the matter in the atom containing the electron and eventually to all the matter in all the atoms of the body containing the atom containing the electron which was struck.

For if this momentum gained could not, in some way, be transferred from the very small piece of the body initially struck by a piece of the aether to the entire body, then the phenomenon of gravity could not exist in the way that I suppose. For little pieces of the body would be knocked off the body without affecting the total momentum of the body itself.

And if we start thinking about this problem of how momentum could be transferred one can draw some very interesting conclusions, based once again upon the guidance that Newton has given us.

But what I would like the reader to consider, first and foremost, is that whatever the process is, it could well work so quickly that millions of these processes could take place in the time that it takes a light wave to vibrate once.

Is it possible for the sun to be virtually transparent as far as gravity is concerned?

All of us are familiar with wind, which is the movement of air. In particular, we know that when the wind blows, it can push things along with it. My theory of gravity is in some ways like the wind. I believe that throughout the universe there are small particles of pure matter moving at incredible speeds, at least 83,000 times faster than light. These small particles of pure matter are incredibly dense.

One of our current theories, called the big bang theory, postulates that at some time in the distant past everything in our universe was compressed into a tiny ball. I personally don't believe in that theory. Nonetheless, reputable scientists all over the world agree that most of what we think of as a solid body is merely empty space. We know that bodies are composed of atoms of different kinds, such as carbon, oxygen, iron, uranium etc. And we also know from work that Rutherford and many others have done that most of the matter within an atom is composed of very dense smaller particles called protons, neutrons, electrons etc. which occupy but a tiny volume of the total volume of each atom. We also know that some of these small dense bodies, such as protons and electrons, have electrical properties which we can calculate but cannot as yet explain.

Current scientific theory also suggests that the larger bodies such as protons, neutrons and electrons are made up of even smaller particles called quarks, some of which have extremely small lifetimes and then break down into other types of quarks. Our current theories on quarks predict that many of these bodies have electrical charges associated with them. As I indicated in proposition 1, energy is matter in motion. And as I indicated in proposition 7, all forces such as magnetism and electrical charges are not properties of matter itself but are rather the result of particles of matter in the aether acting upon much larger bodies of matter in some as yet unimagined way to produce the properties of electrical charges. The fact that quarks behave in a way that can be explained by assuming that they have certain electrical charges indicates to me that quarks themselves must be very large bodies composed of much smaller bodies interacting with the aether to produce the properties of electrical charges.

So even the smallest bodies currently theorized, quarks, must themselves be composed of even smaller particles moving together in some way, but certainly separated from each other by significant distances relative to the sizes of the smaller particles involved.

Think of a building built with bricks. Most of the building is empty space except for the walls which are built with fairly dense bricks. If one were to work out the average density of the building, by taking the total weight of the bricks and dividing it by the volume of the building, the density of the building would be very much less than the density of the bricks themselves. And if we were to now look at a brick and break it down into it's atoms, we would find that the density of the component atoms in the molecules of the brick would be higher than the brick itself, for there is a lot of empty space in a brick, not only as a result of the firing process, but also from the way in which atoms and molecules arrange themselves in crystalline structures. And if we start breaking down an atom into its protons, neutrons and electrons, we find that the density of the protons, neutrons and electrons are very much higher than the density of the atoms themselves. And if we then start breaking down protons, electrons etc. into their component particles, called quarks, we will once again find that the density of matter in a quark is very much higher than the density of matter in a proton. And I feel that each quark is in turn made up of even smaller, more dense particles which could in turn once again themselves be made up of even smaller, more dense particles.

The main thing that I am trying to explain is that as we go from large to smaller and smaller particles, what seems to be happening is that at each level, i.e. from the house to the brick, from the brick to the molecule, from molecules and atoms to protons, neutrons and electrons, from protons to quarks etc. as the particles get smaller, their density increases and the total volume of empty space in relation to the volume of the particles themselves increases.

The proponents of the big bang theory find no difficulty in imagining that if we could get rid of all the empty space in the universe and just concentrate all the pure matter together, we could fit all of the matter in the universe into a tiny ball. While I do not agree with the big bang theory, I do feel that as we keep on getting to smaller and smaller particles, the density of those particles increases to magnitudes that I cannot easily visualize.

What I have proposed, in proposition 5, is that as we keep on getting to smaller and smaller particles, we will find that they keep on getting denser and denser until at some point we will end up with particles of pure matter which are extremely small and dense. And once you get to pure matter, if you then try to divide those particles up then you just end up with smaller pieces of the same pure matter with the same density. And, because any piece of pure matter occupies a certain volume of space that cannot be occupied by any other piece of matter at the same time, it cannot be compressed.

I have no idea of how dense this pure matter is but I am sure that scientists in future generations will find ways to measure the density of pure matter. While I do not agree with the big bang theory, I am quite prepared to accept that, even if we cannot squash our entire universe into a tiny ball, it might at least be quite possible to squash our sun, which is a ball 880,000 miles in diameter, into a ball the size of a walnut. And if this is so, it should at then be possible to squash our whole Earth up into a ball the size of a grain of fine sand, one hundredth of an inch in diameter.

Now lets suppose that we have our sun, consisting of all the matter in it, squashed up into a tiny ball the size of a walnut. Now the sun weighs about 2.0E+30 kgs, (a 2 followed by thirty zeroes). And each gm. mole of the sun contains 6E+23 molecules. So, in terms of protons, neutrons and electrons alone, we are probably talking in the order of E+57 particles. But each of these is made up of even smaller particles. So try to imagine how small and dense each of these little pieces of matter must be if we can squash them all into a walnut. If the sun was squashed up into a ball the size of a walnut, it's density would be about

2.0E+32 times as much as water. So the density of pure matter would be at least 2.0E+32 times as much as water.

Now lets suppose we let the sun, which was compressed to the size of a walnut, grow back to it's normal size of 880,000 miles in diameter. What I want you to understand is that when we expand the sun from the size of a walnut back to it's normal diameter, the pure matter in the sun does not expand and grow less dense. What really happens is that the pieces of pure matter just move further apart from each other. The total volume of the more than 1.0 E+57 particles of pure matter in the sun would still remain less than the volume of a walnut even though they have now moved into a volume of over three hundred and fifty thousand million million cubic miles. In expanding from an inch to 880,000 miles across, each little piece of pure matter would move at least fifty thousand million times further apart from it's neighbor than it previously was. The total volume in which the walnut sized volume of matter would be distributed would have increased by more than one hundred and seventy million million million million times.

Can you imagine that if we could somehow "see" each of the particles of pure matter in the sun, they would be so small and far apart from each other that the sun would appear to be virtually transparent with these incredibly dense little particles of matter spread apart so thinly that the chances of any one of the pieces of pure matter obscuring another from our view would be extremely low. Try to imagine half a cubic inch of pure matter spread evenly in an empty space of three hundred and fifty thousand million million cubic miles of empty space. There is just so much volume present for each little piece of matter that the chances of any one piece of matter obscuring our view of any other piece is extremely low.

And if we think of our Earth, the chance of one piece of pure matter in our Earth obscuring another in the Earth from our view would be even less than that in the sun due to the fact that the Earth is 110 smaller in diameter than the sun.

The average skeptic may well ask why, then, if matter is as thinly spread as I suggest, can't we look through the Earth and see the stars on the other side of the Earth. The reason is that light is an electromagnetic phenomenon that is generated by atoms and molecules. Light moves in waves through the aether and the waves are so large that they are trapped by atoms and molecules, irrespective of how much empty space there really is between the smaller parts within atoms and molecules. Imagine a sieve. Although fine sand may pass through the sieve, a large stone cannot. When we think of trying to look through the Earth with light, its like trying to push your hand through a sieve. Light is just too big and won't fit through. The smallest parts of a wave of light are just too big to fit through the average space between atoms for very long before being stopped.

Now lets imagine an extremely small particle of pure matter in the aether traveling through empty space and reaching our sun. Relative to the size of this particle of pure matter, the spaces between the pieces of pure matter in the sun are enormous. To this piece of pure matter, the sun will truly appear to be virtually transparent and the chances are extremely high that it would be able to travel right through the sun in a perfectly straight line without ever hitting anything in the sun.

Now let us imagine that we have a great wind of tiny particles moving from deep space in the direction of a straight line from deep space, through the sun and then directly to the Earth. Most of the particles will pass straight through the sun and reach us and most of them will pass straight through the Earth without hitting anything.

But a small fraction of the particles passing through the sun will actually hit matter in the sun and be reflected away and will not reach the Earth. And because the matter in the sun is spread apart so widely, the probability of any piece of matter in the sun being hit by any particle of matter moving in the direction of a straight line from that piece of matter to the Earth would not depend upon the presence of any other matter in the sun, for the probability of any two pieces of matter in the sun being in a direct straight line with any particle of matter the Earth are extremely small.

And this explains the first of Newton's questions as to how can the gravity from a body pass through another body without hindrance.

Why does gravity depend upon the total mass of a body rather than it's cross sectional area?

The second phenomenon about gravity that puzzled Newton and which needs to be explained is why does gravity depend upon the total mass of a body rather than on the cross sectional area projected by the body. This is a more complicated question to answer and I will have to give quite a few analogies to try to explain the reason in non scientific terms.

Think of a sailing ship being blown along by the wind. What happens is that the sail catches a certain portion of the wind and transfers the momentum of the air molecules to momentum in the ship. And if you double the area of the sails in such a way that the new sail area added does not interfere with the original sail area, you will double the force applied to the ship by the wind.

In any kind of wind effect involving a stream of small particles colliding with a larger body and imparting momentum to that body, the effect will always be in proportion to the area of the large body projected at right angles to the direction of the stream of smaller particles.

Newton showed, from measurements of their orbits, that all of the planets, including the Earth, as well as Jupiter and Saturn obeyed the same law of gravity, namely that the force of gravity upon all of the planets is in proportion to the mass of the planets. The problem that Newton clearly saw was that this force of gravity was clearly not in proportion to the cross sectional areas that each of the planets projected towards the sun.

The reason for this is simple geometry. All the planets may be regarded very nearly as spheres, or balls. The volume of a sphere is proportional to the cube of it's radius. The cross sectional area of a sphere, i.e. the area projected by a sphere upon another body, is a circle and it's area is proportional to the square of the radius. This means that if you double the radius of a sphere, its volume, or the amount of matter that it could contain, will increase by eight times but the cross sectional area will increase by only four times.

The Earth is about 8,000 miles in diameter whereas Jupiter is about 87,000 miles or nearly eleven times as much as the Earth. The volume of Jupiter is over a thousand times that of the Earth whereas the cross sectional area is just under 120 times as much. Newton proved, by measurements made of the orbits of the planets, that the force of gravity between the sun and the planets, such as Jupiter and the Earth, is proportional to the masses of the bodies and not to the cross sectional area of the bodies.

How then can we explain gravity, which depends upon the mass of bodies, with any theory involving a type of wind action where the force of such a wind action would undoubtedly depend upon the cross sectional areas of the bodies.

Newton was clearly thinking in the correct direction when he got close to postulating that matter was made up of atoms. The problem that Newton had, in his day, was that nobody realized how much empty space there is between particles of matter. People in Newton's time did not even understand the concepts of chemical elements and compounds.

Now think of a bag of half inch diameter glass marbles. Suppose we drop the bag of marbles on an open floor and the marbles spread around in all directions on the floor. If we look down at the marbles on the floor what we will see is the cross sectional projection of each marble towards us. Each marble will look like a circle of half inch diameter. As long as the marbles are separated from each other by a sufficient distance, we will be able to see all the marbles.

Now the interesting thing to think about is that because all the marbles are of the same size, the total cross sectional area of the marbles that we will see is proportional to the number of the marbles. But because the marbles are of all of the same size the total mass of all the marbles which we see is also proportional to the number of marbles and hence to the total cross sectional area of the marbles. So if we drop three times as many marbles on the floor, the total mass of the marbles will increase by three times and the total cross sectional area of the marbles visible to us will also increase by three times.

Hence in any situation where all pieces of matter are composed of spheres of exactly the same size and density and the spheres are separated from each other by a sufficient distance so that no one sphere blocks the view of another, the total cross sectional area projected by the spheres will be proportional to the masses of the spheres. Under such circumstances, any phenomenon which was proportional to the total cross sectional area of the spheres would, coincidentally, also be proportional to the total mass of the spheres.

Now let us suppose that we also had a second bag of one inch diameter glass marbles of the same mass as the bag of half inch marbles. Now each one inch marble has as much mass as eight half inch marbles but only has a cross sectional area of four times as much as a half inch marble.

If we dropped these one inch marbles on a floor by themselves, then the same logic which applied to the half inch marbles will also apply to the one inch marbles. The total cross sectional area of the marbles will be proportional to the number of marbles and the total mass of the marbles will also be proportional to the number of marbles and hence to the total cross sectional area of the marbles. The difference, of course, is that the ratio of mass to the total cross sectional area of the one inch marbles will be twice as much as the ratio of the mass to the total cross sectional area of the half inch marbles.

Now what would happen if we mixed the two bags of marbles together and then dropped them all on the floor? The marbles would be mixed up and scattered randomly upon the floor. What can we then say about the ratio of the total cross sectional area of the marbles to the mass of the marbles? One's first impression might be that we cannot say much. And certainly with the small number of marbles in the bag this might be true.

If you start thinking about this a little more, however, you will realize that we can say a great deal if the marbles were evenly scattered and distributed in such a way that for any large area of the floor that we measured there were always eight times as many small marbles in that area as large marbles.

In such a case, provided that we always selected a sample of marbles such that for every big marble selected we also selected eight small ones, then for any size sample that we selected, the total cross sectional area of the marbles would be proportional to the total number of big marbles and hence to the total mass of all the marbles.

Now if instead of just a bag of mixed marbles, we had a billion big marbles and eight billion small marbles, all mixed up equally and spread out thinly over a vast volume of space, and we had to randomly select any sample of a million marbles at a time, then by the laws of probability, the probability is extremely high that the sample that we select randomly will contain marbles very nearly in the ratio of eight little marbles for every big marble. And if instead of a billion big marbles, we had a billion, billion big marbles and eight times as many little marbles, by the laws of probability if we had to select a sample of a billion marbles, the probability that any sample selected would contain eight little marbles for every big marble is even higher than before.

And as the total number of marbles in the mixture increases to infinity always in the ratio of eight little marbles to one big marble, try to visualize that whatever size sample of marbles that we select above a trillion marbles would contain, as closely as we could measure, the same ratio of big to little marbles, namely one big one for every eight little ones.

And if you then think of this infinitely large mixture of marbles, if you cannot select less than a trillion marbles at a time, whatever size sample you selected would have the same ratio of big to little marbles. Under such circumstances, therefore, the ratio of the total cross sectional areas of the marbles to the masses of the marbles of any size sample that you could select would remain the same.

And if you think of this problem a little further, it is clear that the same conclusion could be drawn whether we had eight little marbles for one big marble or a hundred little marbles for every big marble. In fact, for any infinitely large mixture of two different sizes of marbles, mixed together uniformly throughout all the marbles, in any given ratio of big to little marbles, then the ratio of the total cross sectional areas of the marbles to the masses of the marbles would be the same in any size sample taken above a certain minimum size. This ratio would depend upon the ratio of big to little marbles.

And if you think of the problem a little further, you will realize that if we in fact mixed in a third size marble in a fixed proportion to our other two size marbles, then although the ratio of the total cross sectional areas of the marbles to the masses of the marbles would be different to when we only had two different size marbles, nonetheless in the new mixture of three different size marbles, in any size sample the ratio of the total cross sectional areas of the marbles to the masses of all the marbles would be a constant for any similar size sample taken.

Similarly if we add a fourth and fifth different size of marble to our mixture, the rule would once again hold that in any size sample the ratio of the total cross sectional areas of the marbles to the masses of all the marbles would be a different constant for any size sample taken.

And if we were to increase the number of different sizes of marbles in our sample to any number that we choose, and were to mix in marbles from each size of marble to the others in any ratio we choose, then provided that the mixture had enough marbles and was equally well mixed throughout the universe, then the same rule would once again hold in that for any reasonable size sample of marbles from this mixture the ratio of the total cross sectional areas of the marbles to the masses of all the marbles selected would be a constant for any size sample taken. The ratio of total cross sectional area to total mass would obviously depend upon the distribution curve of the number of marbles of each of the sizes which would be constant throughout the universe.

So far we have considered the case only of a uniform distribution of different size spherical particles. If, instead of considering a mixture of large and small marbles, we were instead to consider a mixture of one size of marble with cubes all of one size, one would once again be able to show that the projected cross sectional area for the mixture was proportional to the projected cross sectional areas of cubes randomly aligned in all directions. And one could, by the same argument show that one could have mixtures of all sorts of different shaped particles and still obtain the situation that for any large sample taken, the cross sectional area projected in any direction would be the same as in any other direction and proportional to the mass.

This almost seems counter intuitive due to our everyday experience of dealing with small samples of objects. Think of an everyday experience of dealing with objects of different shapes, such as moving a pile of rocks. There are so few items present that each sample of rocks we take from a pile could be noticeably different from another. The thing to understand is that as the total number of particles in the universe being sampled increases to infinity and as the minimum number of particles in the smallest sample allowed increases, the probability of including all possible shapes and sizes of particles in the sample increases. Once the numbers are large enough, the probability that any one sample is very similar to another sample of the same size is extremely high.

To get a better idea of what I mean, think of another case similar to a pile of rocks but where there are far more individual rocks in the pile. Think of a tropical beach covered with thousands of tons of fine beach sand. Beach sand is formed by water pounding larger objects, such as shells etc. into fine particles, in a similar way to which I imagine particles of matter in space being pounded by even smaller particles of matter in the aether. If one examines beach sand under a microscope, the grains of sand are by no means spherical and are not even all of the same material. But on a given beach, where the conditions of wind and water are the same, it is remarkable how consistent in sizes the particles of sand really are. If one were to take two buckets of the beach sand and dry the sand and screen each of buckets of sand through a series of different size sieves, one would find a remarkable consistency in the distribution of different size particles of sand in both buckets. And even though there were all sorts of different shapes of sand present, such as long flat pieces or short round pieces, for any type of sand there would even be a consistency in the shapes depending upon the types of material present and their crystalline structures.

So if we could assume that in nature all matter is split up into large numbers of very small particles of various sizes but always such that the distribution curve of the different sizes of particles in the mixture was a constant throughout a certain region of the universe, then within that region of the universe if we were to select any sample of a large number of particles, the ratio of the masses of all the particles to the cross sectional areas of the particles in the sample would be a constant which would merely depend upon the distribution curve of the different size particles in that section of the universe.

Now think about what we now know about matter in our universe which Newton did not know. We know that all matter is made up of atoms. We further know that all atoms are made up of protons, neutrons, electrons etc. and that what distinguishes one element from another are the number of protons in the nucleus and what distinguishes an isotope of an element from a different isotope of the same element are the number of neutrons in the nucleus. Furthermore, it appears as if all protons are very nearly alike in size to each other. Similarly, all electrons are all very nearly equal in size to each other but much smaller than the size of protons. Furthermore, there seems to be a very simple relationship that for every proton there is one electron. So even at this crude stage of dealing with just protons, neutrons and electrons, we know that for every proton there is an electron and that every neutron could be considered to be a proton and an electron bound together. And we also know that the protons tend to be of one size, electrons of another size and a neutron roughly the mass of a proton and an electron. So even at this crude stage of knowledge that we have today, one can already start seeing that if we just considered matter to be made up very roughly of electrons, protons and neutrons, we have the interesting situation that we could consider the universe to be made up of just three different sizes of marbles mixed together in a very definite and fixed ratio throughout the universe. And if one considers the total number of atoms that are present in one gram mole of a substance, 6.0E+23, you can see that we are dealing with so many particles that even the smallest samples of matter that we can weigh have more than a trillion particles present.

But protons, neutrons, electrons etc. are most definitely made of even smaller particles, which move together in ways that cause electrical charges. And since the electrical charges of protons, electrons and the various kinds of quarks postulated are very constant for each type of particle, we can be reasonably certain that this is as a result of the smaller particles of which they are composed also having very definite sizes and ratios of sizes.

And if it is the case that all particles of matter in the universe tend to combine together into larger groups of particles at various levels of sizes and in fixed ratios, then ultimately, the ratio of the masses of all the particles to their cross sectional areas will be a constant for any large size sample of matter that we can measure. This constant will depend upon the distribution curve of the sizes and shapes of all the different size particles in the sample.

The density of a substance gives us an idea of how much matter there is per unit volume of a substance. The higher the density, the more matter there is per unit volume and the "heavier" the material feels. I would like to define a new type of measure which gives us a measure of the amount of mass in a substance relative to the cross sectional area that that substance projects in any direction in the aether. And I would like this measure to be as analogous as possible to density. The higher the density of a substance, the more matter there is per unit of volume. For this reason, I feel that the unit of measure should be the mass of a substance divided by it's cross sectional area, and not the other way around.

Now, as mentioned above, it is my feeling that when dealing with any large number of particles of regular matter, the total mass of the particles divided by the total cross sectional areas of all the particles will be the same over a large volume of the universe.

Let us define now define a symbol, Θ, the universal ratio of mass to cross sectional area of a regular body, as

$$\Theta = \frac{M}{X}$$

or

$$M = \Theta.X$$

where X is the sum of the cross sectional areas of all the ultimate little particles of matter in a body, viewed from any particular direction, and M is the total mass of all the particles of matter in that body. It is my contention that for any large body, consisting of more than a few billion atoms, chosen anywhere over a large volume of the universe, then the mass to cross sectional area of all the component pieces of matter of the body would be Θ.

If this is so, then in any situation where any action is proportional to the cross sectional area projected by all the particles in a body, the action would therefore also be proportional to the total mass of all the particles in the body.

The error in Newton's thinking was in thinking of the cross sectional area of a body, such as a planet, as being the cross sectional area of the one large body, the planet, itself. This is wrong. For when one considers the cross sectional areas of all the particles which a large body such as a planet is composed of, the total cross sectional area of all the component particles is, as has been explained above, proportional to the mass of all the particles. And hence the force of gravity, which we think of as being proportional to the mass of a body, is in truth really proportional to the total cross sectional areas of all the component pieces of matter that the body is composed of. In the kind of bodies which we have measured to date, namely the sun and planets, this is proportional to the masses of the bodies.

And this explains the second of Newton's questions as to why is the gravity from a body proportional to its mass and not it's cross sectional area. For gravity is, in fact, actually proportional to the cross sectional areas of the component particles of matter. The total mass of a body is coincidentally, for the reasons explained above, also proportional to the total cross sectional areas of the component particles of matter in a body. And therefore, gravity is also proportional to the total mass of a body.

Is it logical to assume that all particles of matter in the universe tend to combine together into larger groups of particles of various size in certain fixed ratios?

This question is fundamental to my theory of gravity. For if particles of various sizes do not occur throughout the universe mixed according to a constant distribution curve, then one cannot assume that the ratio of the mass of a body is proportional to the cross sectional areas of all the particles of matter in the body.

This is a problem that I have wrestled with, without realizing it, for decades. The answer finally came to me when I was looking at a particularly unusual snow storm in Connecticut in December 2003. It was about ten o'clock in the morning and the snow was falling heavily and in a very unusual manner. The flakes of snow that were falling were enormous, each averaging easily over an inch in length. As I stood watching the snow falling, I started wondering why the flakes were so large. And as I looked more carefully, I could see that the bigger flakes were falling at a different speed to smaller flakes and were combining with smaller flakes to form even bigger flakes. It slowly became clear to me that the conditions of temperature, humidity and almost total absence of wind had combined to form most unusual conditions that favored the formation of very big flakes. After about ten minutes something changed because the snow gradually changed into much smaller flakes and I have never again seen such large snow flakes as on that day.

And then I remembered a hail storm that I had seen in Johannesburg, South Africa, at about 5:30 p.m. one afternoon in 1984 or 1985. The hail falling was most unusual in that the hail stones were easily the size of golf balls. And when I think back on all the hail storms that I have seen, for any particular storm the size of the hail stones fell within a very narrow range of sizes. If, for instance, the size of the average hail stone in a storm is the size of a pea, one will find a few larger ones but hardly any larger than an inch in diameter or smaller than a tenth of an inch in diameter.

And I have since then observed that in a rain storm, at any particular time in a storm the rain drops falling tend to all be within a very narrow range of sizes. I have found the easiest way to get an idea of the sizes of raindrops is to view the rain drops falling on a car windshield and observing how they spread out before the windscreen wiper clears them from the glass. When this is done it becomes quite clear that even though the size of rain drops may vary from storm to storm, during a period of a few minutes it is remarkable how consistent the drops are in size. One can get very fine drops, such as when it is foggy or misty, or much larger drops in summer when rain falls more heavily.

And when you start to think about why snow flakes or hail stones or rain drops vary in size, one begins to realize that the sizes depend upon a number of things such as the temperature gradient through which the water falls, the air pressure, the humidity, the height from which the water falls, the wind and many other things. But what actually happens is that for any one combination of all of these variables, one will find that the individual bodies of snow or hail or rain that fall are remarkably similar to each other.

And the reason for this is that there are obviously competing factors involved, some of which tend to make the bodies larger and others which tend to make the bodies smaller. Depending upon the competing factors, the snow flakes or hail stones or rain drops or grains of sand on a beach will all tend to either be larger or smaller depending upon the factors involved.

And I think that the same principles are involved in the formation of any type of body which is randomly composed of other smaller bodies. Their sizes will fall within a certain range of depending upon the competing conditions acting upon those bodies, some of which tend to make them bigger and others which tend to make them smaller. And if the conditions which affect the size of a body are very uniform throughout a large volume of the universe, then one would expect bodies of similar sizes to form throughout that volume of the universe.

Think of some of the things that we know about bodies, both larger and smaller than raindrops.

On the large size, there are galaxies. Although galaxies are by no means all the same size, their sizes definitely fall within a certain range of sizes and, from what I can tell from photographs taken by the Hubble telescope, this same range of sizes appears to be uniform to as far as we can see into the universe, both with respect to time as well as position.

There is also a definite size distribution of stars where most stars fall into a fairly narrow range of sizes. Our own sun is a fairly normal star. There is a small percentage of much larger stars but the probability of finding a much larger star decreases rapidly as the size increases. And one can understand the reason for this. Clearly, somewhere on the size scale there is something which ensures that as a star continually increases in size, one will get to a point where things begin to get unstable and something probably happens which will cause the star to shed excess material, such as by overheating and blowing up. Similarly, if a star is too small, there will not be enough mass and pressure to cause the fusion reactions present in stars. One then will end up with a body similar to our planets where the amount of energy generated by internal fusion reactions is very small.

And on the really small scale, if we start thinking about molecules and atoms, one will realize once again that there are limiting factors which determine the size of molecules and atoms. The limiting factors in the case of molecules are things such as temperature and pressure. If one thinks about a simple molecule such as water, for a given pressure, the molecules change states from solids to liquids to gasses at very constant temperatures. We take this for granted, but when you start to think about it, it obviously means that there is a tremendous degree of uniformity between different molecules of water. We also know when molecules react with each other in chemical reactions, there is once again, for any kind of reaction, a very exact and definite point of chemical equilibrium.

If we start thinking about atoms themselves, once again we find very definite limiting factors to their sizes. First of all, atoms are composed of protons, neutrons, electrons and other particles each of which tends to be uniform in size. One of the limiting factors appears to be that at the temperatures and pressures that we experiment with, atoms become quite unstable once one starts getting to more than a hundred protons in an atom. And even for one particular element, such as carbon, one find that although there are different isotopes, these all decay with different half lives which are very precise and constant for any isotope. This once again suggests certain limiting factors determining the half lives of different elements and their isotopes.

Furthermore, we know from spectral analysis, that different types of elements, such as hydrogen, carbon, iron etc. emit light with very definite peaks of intensities at certain wave lengths depending upon the element involved. We also know, when account is taken of the red shift which takes place which is a function of distance traveled, that elements from very distant stars appear to be emitting radiation with the same spectral patterns that are emitted from stars close to us. This indicates that whatever the limiting factors are within atoms and molecules that determines their spectral patterns, they are very constant thoughout our known universe not only with respect to position but also with respect to time. This once again suggests that we are dealing with a situation where the sizes of atoms and their constituent parts are very constant and uniform throughout our known universe both with respect to position as well as time.

Current theory suggests that protons, neutrons and electrons are themselves made up of a small number of different types of quarks. If this is indeed so, then even at the level of quarks, which are the smallest particles that we have any experimental evidence for, we have already found evidence that the entire universe might be built up of no more than a dozen or so different types of quarks, each with very definite and fairly consistent properties. And since the evidence which I have listed above suggests that atoms are very constant throughout our universe, then it is difficult to imagine how the component quarks could be different in different parts of the universe without this resulting in noticeable differences in atoms.

The interesting thing is that current theories merely assume that different types of quarks have different masses and electrical charges without considering the reasons why this is so.

My feeling, as mentioned in my earlier propositions, is that all things that we refer to as properties of matter are not actually properties of matter itself at all but rather the result of interactions between the aether and matter which appears to give matter the properties that it has.

I feel that at the very lowest level, when one is dealing with particles of pure matter which is composed of nothing else than pure matter, that there are no properties other than inertia involved. These small particles interact with other small particles randomly and as a result of collisions are continually splitting up or joining to form larger bodies. My feeling is that there are laws of probability which determine, when one considers vast numbers of particles of pure matter moving at enormous speeds, how they will combine so that large quantities of matter eventually get concentrated in relatively slow moving bodies, such as quarks, protons, neutrons, electrons, atoms, molecules, planets, stars, galaxies, super clusters of galaxies etc. Somewhere along the line of progression from extremely small to large bodies, matter in the aether concentrates itself at some level and moves in such a way as to account for the properties that we call electrical charges and magnetism.

I feel that one of the principles that we can take for granted is that whenever we do find large numbers of bodies that tend to be of very consistent and definite sizes, such as galaxies, stars, snow flakes, molecules, atoms, protons, quarks etc., then this is a very good indication that they themselves must be made up of even smaller particles. For there is nothing magical in the universe and something in the end must determine the sizes of these bodies. They do not just happen by accident. In the end, in each case it will be found that whenever bodies are of a similar size, such as galaxies or stars, hail stones or snow flakes, atoms or protons or quarks, it will always be due to even smaller particles combining together randomly but always constrained by the laws of probability and competing factors which strongly favor the formation of certain size bodies over the formation of similar bodies of either larger or smaller sizes.

The task of future generations is to determine these factors which cause things such as protons, neutrons, quarks, etc. to form at particular sizes and to explain what causes electrical charges, magnetism and electromagnetic radiation.

But in the meantime, I feel that it is quite logical to assume that all particles of matter in the universe tend to combine together into larger groups of particles of various size in certain fixed ratios.

In one way, the best proof that we have of this is that the force of gravity between two bodies is indeed proportional to their masses, at least in bodies of the sizes that we have measured.

How is it possible that the aether offers no resistance to the planets in their orbits about the sun that would cause them to slow down?

Newton first posed this question and in doing so made the same incorrect assumption that many other scientists since then have made. Newton was well aware, as he explained in his description of experiments that he performed with pendulums, that the air offers resistance to the motion of moving bodies. Newton, in fact, developed ways to estimate the effects of air resistance upon pendulums. Newton then made the assumption that the aether must somehow be similar to air in that when one considers bodies moving through the aether, the bodies are moving faster than the aether itself and are somehow pushing the aether aside as they move through it.

For air resistance is caused by a large body having to push aside much smaller particles of air in order to be able to pass through the air. In doing so, the large body has to impart some of it's momentum to the air particles to move those particles out of the way. The same kind of thing happens when a ship moves through water. The water has to be physically moved out of the way for a ship to pass through it.

The aether itself is not at all like air or water.

First of all, the particles of matter in the aether move very much faster than larger bodies such as planets and stars. If we consider the Earth moving around the sun, it moves around the sun at a speed of about 30 Kms/sec. As I showed in Section 1, gravity moves at a speed of at least 2.5E+10 kms/sec or about a thousand million times faster than the Earth around the sun. So, when considering a body such as the Earth moving through the aether, in terms of speeds involved the situation is very different from a bullet speeding through the air. In the case of the bullet, it is the bullet that is moving very much faster than the air and it has to push the air aside to pass through the air. In the case of a planet, it is moving so much slower than the particles of the aether that it does not have a chance of catching up to even the slowest piece of aether. In fact, unlike the bullet in air, it is the aether that is moving though the planet rather than the planet moving through the aether. And unlike a bullet in the air, where the bullet is moving so much faster than the air, in the case of a planet, it is the aether that is moving so much faster than the planet and it is the aether that is moving through the planet.

As mentioned previously, the aether is composed of such tiny particles of matter that to them even something as massive as the sun appears virtually transparent and most of the aether can move completely through even something as large as the sun without colliding with anything. This is quite different from a pendulum moving through the air in that the air particles cannot move through the pendulum at all. If they could, the resistance of air to the pendulum would be very much less than it is.

The reason, therefore, that the aether offers no resistance to planets moving around the sun is because the aether moves very much faster than the planets and because the aether can move through the planets with very little resistance.

If the aether moves so much faster than the planets, why does it not push the planets?

The average skeptic may well ask why then, if the aether is as I suggest, would the aether then not push the planets along in very much the same way that a leaf or piece of paper is blown through the air by the wind.

The answer is that the aether can indeed push the planets. And when it does push the planets, we call this effect gravity. For the aether is continually pushing each of the planets towards the sun and that is what keeps the planets in their orbits about the sun. So rather than slowing the planets down, the aether is in fact what is providing the force and energy to keep the planets moving in their orbits around the sun.

On a calm day, when there is no wind, a leaf is not blown through the air. It merely remains at rest. It is only when there is a lack of equilibrium in the air, due to pressure differences, that the leaf is blown through the air. In a similar way, it is only when there is a lack of equilibrium in the aether that the aether will push us through space with a force that we call gravity.

A description of the mechanism causing gravity.

As mentioned earlier, it is my feeling that the aether is composed of a vast number of small particles of matter moving randomly through space at velocities at least 80,000 times greater than that of light.

As I also mentioned earlier, it is my feeling that this motion is very uniform in all directions such that if one were to take a unit of area then the total momentum of all particles moving through that unit of area perpendicular to the area would, in the absence of any large bodies of matter, be the same irrespective of which direction the plane containing the unit of area faced. Furthermore, for any unit of area on a plane, the total momentum passing thought that plane in a second in the one direction perpendicular to the plane would be the same as the momentum of the matter moving through the same unit of area in the opposite perpendicular direction.

I will need to refer to directions and areas repeatedly in this section. I would like to define the following terms with the meanings that I will use for them in this section.

In discussing gravity, one needs to draw a distinction between two types of matter. The one type of matter consists of small particles of the aether moving at speeds very much faster than light. I will refer to these particles of matter as particles of the aether. The other type of matter consists of particles of matter which form part of much larger and slower moving bodies such as atoms and molecules, either at rest or moving very slowly with respect to the velocity of light. I will call these bodies large bodies or bodies of atomic matter.

By a unit of area I mean a unit of area on a flat plane or on a spherical surface where the radius of the sphere is so large in comparison to the size of the unit of area that for all intents and purposes the unit of area can be considered to be on a flat plane touching, or tangent to, the sphere at a point which lies within the unit of area.

If one thinks of a piece of paper it has two faces, or sides. If one is looking at the piece of paper from one side, we would call the one side the front and the other side the back. If one imagines shooting a bullet through the paper, the one side will be the face of entry and the other side, or opposite side, will be the face of exit.

In the same way, it is important when talking about particles moving through a unit of area to be able to distinguish between the two faces of a unit of area. I will refer to these two faces in one of three ways, as the need arises, as any of the following three ways:
  1. As the front and back of the unit of area. When talking about particles striking or passing through a unit of area, I will call the front face the face of entry of the particles and the back face the face of exit of the particles.

2. As the face of entry and face of exit of the area.
3. As one face of the unit of area and the opposite face of the unit of area.

When referring to the direction of a unit of area I will generally refer to the direction of a straight line drawn perpendicular to the plane containing the unit of area.

The total momentum of matter in the aether moving through a unit of area in a second would be the sum of the masses of all pieces of matter times their velocities that enters one face of the unit of area and leaves through the opposite face. As explained earlier, there would be an equal and opposite total momentum of matter in the aether moving in the opposite direction through that same unit of area in the same time entering from the opposite face and leaving from the first face.

The total perpendicular component of momentum of matter in the aether that enters through one face of a unit of area in a second would thus be the sum of the masses of all pieces of matter times the component of their velocities in the direction of a line perpendicular to the plane containing the unit of area. Once again, if the aether is in equilibrium, there would be an equal total perpendicular component of momentum of matter in the aether that enters through the opposite face of the unit of area.

In dealing with matter, it is important to remember that we are dealing with a substance in three dimensions. It is also important, when considering the effect of the aether on a piece of atomic matter, that we bear in mind that this piece of atomic matter is most likely a collection of a very large number of very small pieces of matter all loosely combined together because they share the common properties of all moving at the same velocity in the same direction and all being close to each other.

When dealing with gravity, such as considering the effect of the Earth on a person, we are in fact dealing with the effect of a very large number of tiny pieces of matter in the Earth on a much smaller but still very large number of pieces of matter in a person. More importantly, however, is that in all types of problems concerning gravity, the distances between the two bodies being considered is very much larger than the sizes of the individual little pieces of pure matter in either of the bodies. The effect of this is that when considering the distance between one tiny particle of matter in the first body and one tiny particle of matter in the second body, the distances between the two particles are on average so large in comparison to the sizes of the individual particles that when one squares these distances, it effectively makes no difference what point within each individual particle one uses as the point of measurement. For example, consider the calculation involved in determining the force of gravity between two half inch marbles at a distance of four thousand miles apart using Newton's formula for gravity. It really makes virtually no difference if one measures the distance between the two marbles as being the distance between the two centers of gravity of the two marbles, or the distance between the two closest points in the two marbles, or the distance between the two furthest points in each marble. In terms of differences, one would be talking of 600,000,000 inches plus or minus half an inch all squared. The maximum error in the force involved would be in the order of one in 600,000,000 which is far less than our ability to measure. And as we reduce the size of the marbles, keeping their distance apart the same, the error involved will continue to decrease.

One of the fundamental concepts developed by Newton in the development of the calculus was in considering what happens in situations where the size of something is reduced to zero. If one considers, as Newton did, that each piece of matter attracts another piece of matter with a force proportional to the product of the masses and inversely proportional to the distance between the pieces of matter, then the same rule applies irrespective of how small one makes either piece of matter. If, on the other hand, gravity is not a property of matter itself but is rather an interaction between matter and even smaller pieces of matter in the aether, then one needs to examine if it does matter to some extent on how close to zero we reduce the size of a particle of matter. Clearly, if one reduces the size of the piece of matter being considered to a size smaller than the average size of a particle of matter in the aether, one could clearly not expect Newton's law of gravity to be applicable over a short period of time. Obviously, during the periods where no particles of the aether collide with the particle, it would not experience any force of what we term gravity.

My feeling is that, for all practical purposes, we can still use all the ideas developed by Newton in calculus when considering the problem of gravity. For even in situations where we reduce the size of a body to as small as we desire, that body will still, over a long period of time, experience multiple collisions with other particles in the aether. All that effectively happens, by reducing the size of a body below a certain size is that one must wait a longer period of time for sufficient collisions to take place so that the laws of probability will ensure that the tiny piece of matter would behave in the same way as a much larger piece of matter. Given the speed at which matter in the aether moves, my feeling is that the processes involved in gravity are moving so quickly and at such an exceeding small level, that over any long period of time, such as a second, and over the large distances that form the least distances that we can measure, any effects caused by reducing the size of a particle measured to as small as we desire would become insignificant.

When dealing with gravity, one is really dealing with a combination of masses and the cross sectional areas of those pieces of matter. The effects of gravity really depend upon cross sectional areas of particles more than the masses of the particles. Unfortunately, we have no means readily available at the moment to measure cross sectional areas but we do have very accurate means to measure the cumulative masses. For the purposes of this discussion, I would like to make the following assumptions:

Suppose we choose a unit of area as being the projected cross sectional area of a number of particles that is large enough that we can assume that the formula developed previously between mass and cross sectional area is valid. I discussed previously how, when one considers a large number of particles of atomic matter in a regular body, that the total masses of all the particles will be related to the total cross sectional area of all those particles in the body according to the formula:

$$M = \Theta.X$$

In addition, it was also explained that for any large number of particles, such as those required to have a cross sectional area of our unit of area, the cross sectional area of all the particles viewed in any direction will tend to be the same as the cross sectional area of the particles viewed from any other direction as the number of particles increases. For purposes of this discussion, I will assume that all the particles of matter that combined have a cross sectional area of a unit of area in one direction will have the same area when viewed from any other direction.

If one considers a gas confined in a container it consists of molecules of the gas moving about and colliding with each other in much the same way that I envisage particles of the aether colliding with each other and with very much larger and more slowly moving bodies. One of the most useful measurements that we use when working with gasses is what we term and measure as pressure. After thinking for many years about how to best talk about my concepts of gravity, my feeling is that the best way is to introduce a term that is in many ways analogous to pressure. If one considers a gas under pressure inside a container, then we know that individual molecules of a gas are moving about inside the container. The various molecules are all moving at different speeds, which depends upon their temperatures etc. But what we do know in practice is that the molecules do exert a force on the sides of the container which is, for any amount of gas in a container at a certain temperature, roughly constant over all portions of the container if one ignores the weight of the gas. This pressure is typically measured in pounds per square inch etc. And any other body, significantly larger than the molecules of the gas, such as a coin, placed inside the container will experience an equal force on it by the gas in all directions and will not be moved about inside the container by the gas irrespective of the pressure of the gas.

Consider an infinitely large volume of space occupied only by the aether and with no large atomic bodies present. Consider a point in this large volume and let us consider a flat unit of area passing through that point in any direction. Then if one were to imagine the aether acting upon this unit of area, the aether would exert a force on that unit of area in the same way that a gas does on the wall of a container or, more importantly, something such as a flat coin suspended in the container of gas. In the case of the gas, this pressure is caused by molecules of the gas colliding with the surface of the container and bouncing off it. So, in the same way, if our unit of area was such that matter in the aether could not pass through it, the aether would exert a pressure on the front face in a direction perpendicular to that unit of area. This pressure would be caused by the particles of the aether being reflected by the surface of our unit of area in the same way that a gas exerts a pressure on the inside surface of the container. And just as in the case of a flat coin suspended in a gas in equilibrium under pressure, the aether would, provided that it is in equilibrium, exert an equal and opposite force on the back face of our unit of area. This would result in our unit of area experiencing no net force from the aether. Our unit of area would, however, still be under pressure from the aether in the same way that a flat coin suspended in a gas is under pressure from the gas.

Without concerning ourselves with the exact mechanics of how the aether exerts pressure on large bodies in the aether, which we don't as yet understand, let us define a constant to describe the pressure that the aether exerts on a body in the aether in much the same way that we have defined the pressure in a gas to be the force that a gas exerts per unit of area on any body placed in a gas.

Let us define ϕ to be the total force that the aether would exert on one face of a unit of area in a direction perpendicular to the unit of area under conditions where the aether is equilibrium. ϕ would thus be numerically equivalent to the pressure that the aether exerts on a flat surface. Then as explained previously, the value of ϕ will be numerically the same irrespective of the direction of the unit of area and will be a constant throughout a large volume of the universe in any region of the universe where there are no large particles of atomic matter present to disturb the equilibrium of the aether. This pressure could be thought of as a result of particles of the aether colliding with a body and being reflected from the body with an effective coefficient of restitution of 1.

What we term as the pressure of a gas on a surface is proportional to the change in the perpendicular component of the momentum of all the molecules of gas striking the side of the container. In the same way, the pressure of the aether on any surface is proportional to the change of the perpendicular component of the momentum of all particles that hit that surface of a body and are reflected off it. Let us now consider the rules which would govern how the perpendicular component of the momentum of all particles in the aether passing through or hitting a body could be affected by the presence of another large body in the aether.

Think back to our infinitely large volume of space occupied only by the aether and with no atomic bodies in it. Now let us suppose that we place a body A of atomic matter of mass M_A in this volume of space and let us suppose that body A is stationary in the aether, i.e. the total perpendicular component of momentum of matter in the aether reaching body A in any direction is a constant. As explained previously, provided that the body is small enough, all the particles of matter in this body would be "visible" to a particle of the aether. As also explained, this visible area projected by all the particles of matter in the body will be the same irrespective of the direction from which body A is viewed. Let the total cross sectional area of all the particles of matter in body A viewed from any direction be X_A. As explained previously, the ratio $\dfrac{M_A}{X_A}$ will be a constant throughout a large volume of the universe, indicated by the symbol Θ where

$$\Theta = \frac{M_A}{X_A}.$$

As mentioned in proposition 5, I visualize that in collisions between matter in the aether, the coefficient of restitution will be 1. As I also mentioned, the assumption is that after an impact of a piece of matter in the aether with a piece of matter in a large (atomic) body, the momentum imparted to the large body will eventually be distributed to all the particles of matter comprising the large body.

Now in any direction in a unit of time the total force of matter in the aether striking body A from any direction will be $\phi.X_A$. As explained previously, in the same second of time there would be exactly the same effect in the reverse direction from matter in the aether arriving from the opposite direction. The force exerted in the reverse direction would be $-\phi.X_A$

The net effect of this is that in any direction, the total or net force exerted by the aether upon body A is zero. If one considers the aether itself, the total net change of perpendicular momentum in the aether in either direction is also zero. What has effectively happened is that the amount of aether reflected in each direction exactly balanced the amount of matter in the aether that would have continued on it's passage in a straight line if body A was not present.

The net effect of the large body A being placed by itself in an infinitely large volume of aether is that there is no net change of momentum in body A in any direction and the aether itself is effectively unchanged by the presence of body A in the infinitely large aether. Body A will, therefore, travel through the aether with no net change of momentum, i.e. body A will experience no resistance to it's passage through the aether provided that it moves extremely slowly with respect to the speed of the particles in the aether.

Now let us suppose that we place a second stationary body B of atomic matter of mass M_B in this volume of space. Let us suppose that the distance between the two bodies A and B is R and further suppose that distance R is very large in relation to the dimensions of either body A or B. Let the total cross sectional area of all the particles of matter in body B viewed from any direction be X_B. Then, just as in the case of body A, the ratio $\dfrac{M_B}{X_B}$ will be equal to Θ.

Hence

$$\Theta = \frac{M_A}{X_A} = \frac{M_B}{X_B}$$

In the same way as before, if body A were not present, body B would experience no resistance to it's passage through the aether and the aether would remain unchanged by the presence of body B in the aether.

But the fact that there are now two bodies A and B both present in the aether does introduce a new and complicating factor with respect to both body A and body B.

Consider the effect of body B upon body A. When body A was alone by itself in the aether, it was unaffected by the aether because the total force of matter in the aether striking body A

from any direction in a unit of time was a constant which we worked out to be $\phi . X_A$.

But body B is preventing a very small amount of the aether, which would have reached body A, if body B was not present, from actually reaching body A. The amount of the aether that body B is preventing from reaching body A obviously depends upon how close body B is to body A and how big bodies A and B are. If we think about body A, when it was alone by itself in the aether, the total momentum of aether striking it and being reflected in any direction was exactly balanced by an equal and opposite momentum of aether striking it and being reflected in the opposite direction. But if body B is introduced, body B will reflect a small amount of aether, originally moving towards body B and then A in the direction of the line through B and A. This aether reflected by body B would have reached body A if body B was not present. When body A was alone in the aether without body B being present, the total perpendicular component of momentum of the aether entering the unit of area in the direction from B to A was the same as the total perpendicular component of momentum of the aether entering the opposite face of the unit of area in the direction from A to B. But body B now prevents a certain amount of aether which would have reached the front face of A, in the direction from B to A, from actually reaching body A. The total perpendicular component of the momentum of the aether reaching our unit of area on body A in the direction from B to A will now be less than it was before by the exact amount of the momentum of the aether that body B prevented from reaching body A. But the total perpendicular component of the aether entering the back face of our unit of area, in the direction from A to B, is not affected by the presence of body B and remains the same as before.

The mere presence of body B will, therefore, cause an imbalance in the total perpendicular component of the momentum in the aether reaching our unit of area, passing through body A and perpendicular to line AB. There will, in fact be less aether reaching body A in the direction from B to A than in the direction from A to B. The difference in the amounts of momentum in the two directions will be equal to the amount of momentum in the aether that is prevented by body B from reaching body A. This will cause body A to experience a force acting upon it in the direction from A to B. This force, which we call gravity, has until now incorrectly been thought to be some type of pulling action of body B upon body A. My feeling is that this force of gravity is rather due to a pushing action of the aether pushing body A towards body B.

How the total momentum of the aether depends upon the presence of a distant body

Consider a small piece of atomic matter which projects a unit of area in any direction. Let us choose our unit of area as being the minimum needed so that the error in assuming that the same area is projected in all directions by all the particles of matter comprising the piece of atomic matter is negligible. In view of the fact that we have never been able to detect any weight difference in the smallest piece of matter that we can weigh that is dependent upon it's direction towards the Earth, we may assume that the total volume of our small piece of atomic matter with a unit of area is a very small fraction of a cubic centimeter. In addition, when considering the effects of gravity of bodies, we are typically talking of distances measured in miles or thousands of miles.

So for the purposes of this discussion, the error in assuming that our piece of atomic matter has a spherical shape, as opposed to any other shape, is negligible. Let us now choose a plane in any direction passing through our piece of atomic matter.

Particles of aether will be passing through the front face towards the rear face. Let the total momentum of all the particles passing through the front face, from front to back, in a second be U.

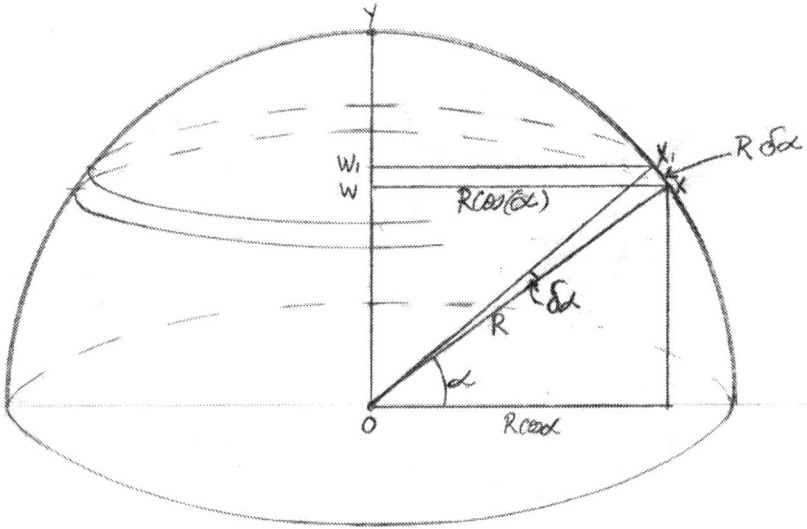

Section 2 - Figure 1

Now consider a hemisphere drawn about our unit of area where the base of the hemisphere lies on the plane of our unit of area and the center of the hemisphere lies within our unit of area as shown in the diagram above. Let O be the center of the hemisphere. O will lie within our unit of area.

Let the radius of this hemisphere be R. Let the radius of the hemisphere be so large in comparison to the dimensions of our unit of area that the specific shape of our unit of area can be ignored, as discussed previously.

Now all the particles of aether passing into the front face of our unit of area must have passed through the surface of our hemisphere on their way from the distant parts of the universe to our unit of area. Clearly the total momentum of all the particles passing through any unit of area on the outside face of the hemisphere will also be U. However, most of these particles will miss our unit of area at the center of the hemisphere. Let U_R be the momentum per unit area of the aether passing through the surface of the hemisphere that will actually reach our unit of area at O. Clearly if the aether is in equilibrium, the value of U_R will be constant over the entire surface of the hemisphere. Clearly, the larger the radius of the hemisphere, the more chance there is that a particle of the aether traveling through a unit of area on the hemisphere will not hit the unit of area at O, the center of the hemisphere. Hence as R increases, U_R will decrease.

Let us now integrate, or add up, the effect of all the aether passing into the surface of our hemisphere. The easiest is probably to simply integrate with respect to an angle drawn from the center of the hemisphere to the surface of the hemisphere.

Let Y be the point on the hemisphere such that line OY is perpendicular to the base of the hemisphere. Consider a plane parallel to the base of our hemisphere passing through our sphere at a distance from the center of the sphere as shown in the diagram above. Let this plane intersect line OY at point W. This parallel plane passing through point W will intersect our hemisphere in a circle with center W. Let X be a point on this circle of intersection on the hemisphere. Let α be the angle between the line drawn through OX and the base of the hemisphere.

The circumference of the circle of intersection between the hemisphere and the plane through points W and X is

$$= 2.\pi.XW$$
$$= 2.\pi.OX.\cos(\alpha)$$
$$= 2.\pi.R.\cos(\alpha)$$

Now let us draw a second parallel plane intersecting our hemisphere at a slightly greater angle $\alpha + d\alpha$. Let this plane intersect the hemisphere at point X_1 as shown such that all points Y, W, O, X and X_1 line on a plane perpendicular to the base. Then

$$\text{distance of arc from } X \text{ to } X_1 = R.d\alpha$$

where $d\alpha$ is vanishingly small. The second plane will once again intersect the hemisphere in a circle with center W_1

Consider the area of the hemisphere lying between these two circles of intersection with centers W and W_1 passing through points X and X_1. This area is in the form of a ring with circumference $2.\pi.R.\cos(\alpha)$ and height $R.d\alpha$.

Hence

$$\text{Area of ring on hemisphere} = 2.\pi.R.\cos(\alpha).R.d\alpha$$
$$= 2.\pi.R^2.\cos(\alpha).d\alpha$$

This ring is perpendicular to line OX when $d\alpha$ is vanishingly small. Hence the total momentum of aether passing through this ring on it's way to our unit of area at point O is:

$$\text{Total momentum through ring} = U_R.\text{Area of Ring}$$
$$= U_R.2.\pi.R^2.\cos(\alpha).d\alpha$$

To get the total momentum of all the aether passing through our hemisphere on it's way to our unit of area we simply integrate all the little rings obtained by varying α from 0 to 90 degrees, i.e. from 0 to $\pi/2$ radians.

Hence, the total momentum of the aether passing through the hemisphere on its way to our unit of area is:

$$= \int_0^{\pi/2} U_R.2.\pi.R^2.\cos(\alpha).d\alpha$$

$$= U_R.2.\pi.R^2. \int_0^{\pi/2} \cos(\alpha).d\alpha$$

$$= U_R.2.\pi.R^2.[\sin(\alpha)]_0^{\pi/2}$$

$$= U_R.2\pi.R^2.(1-0)$$

$$= U_R.2.\pi.R^2$$

This is simply the area of the hemisphere times U_R as we would expect.

Now that we are on the right track, let us investigate the perpendicular component of the momentum reaching the surface of our unit of area.

Using the same figure as above, it is clear that all the particles of the aether passing though the same circular ring on our hemisphere, of center W and angle α, are all striking our unit of area, or the base of the hemisphere, at an angle of α to the base.

The perpendicular component of the momentum of all these same particles through our unit of area at O is thus $\sin(\alpha)$ times the total momentum. Hence the perpendicular component of the total momentum of the aether passing through our unit of area about point O, which originally passed through the ring on the sphere is $U_R.2.\pi.R^2.\cos(\alpha).d\alpha.\sin(\alpha)$

Hence, following a similar line of logic to that which we used to calculate the total momentum, the total perpendicular component of the momentum of the aether passing through the hemisphere on its way to our unit of area is:

$$= \int_0^{\pi/2} U_R.2.\pi.R^2.\cos(\alpha).d\alpha.\sin(\alpha)$$

$$= U_R.2.\pi.R^2. \int_0^{\pi/2} \sin(\alpha).\cos(\alpha).d\alpha$$

$$= U_R.2.\pi.R^2. \left[\frac{\sin^2(\alpha)}{2} \right]_0^{\pi/2}$$

$$= U_R.\pi.R^2. \left[\sin^2(\alpha) \right]_0^{\pi/2}$$

$$= U_R.\pi.R^2.(1^2 - 0)$$

$$= U_R.\pi.R^2$$

We thus arrive at the rather interesting fact that the perpendicular component of the momentum of the aether passing through a unit of area, in one direction, is numerically equal to half of the total momentum of the aether passing through that unit of area.

Previously we defined ϕ to be the total force that the aether, in equilibrium, exerts on one face of a unit of area in a direction perpendicular to the unit of area. This force that the aether exerts on one face of a unit of area is directly proportion to the total perpendicular component of the aether passing through our unit of area. But, we have just shown that the total perpendicular component of the momentum of all particles of the aether passing through our unit of area is $U_R.\pi.R^2$. Hence:

$$\phi \text{ is proportional to } U_R.\pi.R^2$$

The above equation appears more simple than it actually is. I would like to discuss the above equation in a little more detail as I feel that it is always important, when dealing with mathematics, not to fall into the trap of getting so caught up in mathematical manipulations that we forget what we are doing or what the symbols in the mathematical equations mean.

Previously we defined U_R be the momentum per unit area of the aether passing through the surface of a hemisphere that will actually reach our unit of area at the point O. The hemisphere was defined as having a radius of R and the base of the hemisphere in the same plane as our unit of area and the center of the base of the hemisphere being within our unit of area.

In the above equation ϕ and π are constants. Rearranging the above equation gives:

$$U_R \text{ is proportional to } \frac{\phi}{\pi.R^2}$$

Hence it can be seen that if one considers a sphere of any radius centered about our unit of area, the total momentum of the aether passing through a unit of area on the sphere that will also pass through our unit of area at the center of the sphere, namely U_R, is inversely proportional to the square of the radius of the sphere and is directly proportional to the pressure that the aether, in equilibrium, would exert on a unit of area of a surface.

At this point, understanding the above equation, we can now make the final logical conclusion which took me so many years to fully understand.

If, at any point on the surface of the hemisphere, we prevent the aether traveling through a certain area of that hemisphere, then the total perpendicular component of the momentum of the aether reaching our unit of area at O will be reduced by the perpendicular component of the momentum of the aether that would have previously reached the unit of area at O but which now cannot. This means that the total pressure of the aether acting on the unit of area at O will be reduced.

In particular, suppose that we consider one particular area of interest, namely that at point Y on our hemisphere. Going back to our original figure, line OY is perpendicular to the unit of area at O and to the base of the hemisphere. Let us suppose that around point Y we take a small area on the hemisphere, of area ΔX, and prevent all aether from traveling through this little area.

Then the total momentum of aether that will not reach the unit of area at point O is simply $U_R.\Delta X$. But, because of our choice of point Y, all of this aether that would have reached point O would have arrived in a direction perpendicular to the unit of area at O. Hence the total perpendicular component of the momentum of the aether reaching the unit of area at O will be reduced by the amount $U_R.\Delta X$. Hence the total pressure exerted by the aether, or force per unit area, exerted by the aether on the unit of area at O will be reduced in proportion to the reduction of the perpendicular component of momentum, $U_R.\Delta X$. Let this reduction in pressure of the aether be $\Delta\phi$

Then $\Delta\phi$ will be proportional to the reduction of the perpendicular component of momentum, $U_R.\Delta X$, in the same way that the total pressure of the aether, ϕ, is proportional to the total perpendicular component of momentum of the aether passing through a unit of area, $U_R.\pi.R^2$.

Hence

$$\frac{\Delta\phi}{\phi} = \frac{U_R.\Delta X}{U_R.\pi.R^2}$$

$$= \frac{\Delta X}{\pi.R^2}$$

From this it follows that

$$\Delta\phi = \frac{\phi.\Delta X}{\pi.R^2}$$

The above equation is, to my mind, a fundamental equation in gravity and once again need to be discussed in simple words to be fully understood. What this equation says is that if we have a unit of area in the aether at point O and a second point Y in the aether at a distance R from O, such that line OY is perpendicular to the plane containing the unit of area at O, and we then prevent the aether from passing through a small area ΔX at point Y, this area ΔX being parallel to our unit of area at point O, then there will be a net reduction in the force of the aether on the unit of area at O that will amount to exactly $\dfrac{\phi.\Delta X}{\pi.R^2}$. This net reduction of force will be in the direction of the line joining O and Y and will be such that a body at point O would be pushed by the aether in the direction from O to Y.

Calculation of the force of gravity between two bodies.

Let us now consider two atomic bodies, A and B separated by a distance R. Let X_A and X_B be the cross sectional areas projected by the bodies A and B and let M_A and M_B be their respective masses.

Now think about the effect of body B upon body A. The matter in body B will prevent a certain amount of the aether from reaching body A. This will cause a net decrease in the force, or pressure of the aether, acting upon A in the direction from B to A.

We have shown above that for an area ΔX of body B at distance R from body A, the total reduction in the force of the aether per unit of cross sectional area of body A, $\Delta\phi$, will be given by the formula

$$\Delta\phi = \frac{\phi.\Delta X}{\pi.R^2}$$

Since the total cross sectional area of body B is X_B, the total reduction in the force of the aether on body A per unit of cross sectional area of body A will be

$$\text{Total reduction of force per unit of area } A = \frac{\phi.X_B}{\pi.R^2}$$

This assumes that the dimensions of body B are very small in relation to the distance R.

So the total reduction in the force of the aether, F_A, on the entire body A will this be the cross sectional area of body A times the reduction of force per unit of area on A, once again assuming that the dimensions of body A are very small in relation to the distance R.

Hence the total reduction of the force of the aether on body A is:

$$F_A = \left(\frac{\phi.X_B}{\pi.R^2} \right).X_A$$

This reduction of force, F_A, on body A is what we call the force of gravity experienced by body A due to the presence of body B in the aether.

Hence

$$F_A = \frac{\phi.X_A.X_B}{\pi.R^2}$$

Now, as mentioned previously, if both bodies, A and B, are normal atomic bodies of the type that we are familiar with, then their cross sectional areas, X_A and X_B, will be related to their masses according to the equations:

$$X_A = \frac{M_A}{\Theta}$$

and

$$X_B = \frac{M_B}{\Theta}$$

Substituting these two relationships into the equation for the force of gravity acting on body A gives:

$$F_A = \frac{\phi.M_A.M_B}{\pi.\Theta^2.R^2}$$

This resultant force, acting on body A, pushes body A with a force that is proportional to the product of the masses of bodies A and B and inversely to the square of the distances between the bodies. This equation is of exactly the same form as suggested by Newton for gravity.

The important thing to understand is that this resultant force acting on body A is not, as has hitherto been assumed, a pulling force emanating from body B. Body B possesses no magical property of gravity that enables it to attract body A towards it.

As has been shown, the force on body A acting in the direction towards body B is rather a pushing action of the aether upon body A pushing it towards body B. This pushing action is due to the imbalance in the aether caused by body B preventing a certain small amount of the aether from reaching body A. This imbalance results in slightly more aether reaching body A in the direction of A to B than reaches A in the direction from B to A.

Now, in our derivation of the force acting upon body A by the aether due to the presence of body B, there was nothing particularly unique about our choice of bodies A and B other than they were a distance R apart and were stationary with respect to each other and to the aether.

We can use exactly the same arguments as before to calculate the force acting upon body B by the aether due to body A preventing a certain amount of aether from reaching body B.

If we follow this argument we will arrive at:

$$F_B = \frac{\phi.X_A.X_B}{\pi.R^2}$$

and, under the same conditions as before with respect to bodies A and B, we get

$$F_B = \frac{\phi.M_A.M_B}{\pi.\Theta^2.R^2}$$

In this case, the force acting on body B will act in the direction from B to A. As can be seen, this force is numerically the same as the force acting upon body A but acts in the reverse direction.

We could write the above two equations in the following form, taking into account that F_A and F_B act in opposite directions:

$$F_A = -F_B = \frac{\phi.M_A.M_B}{\pi.\Theta^2.R^2}$$

This can be written in the standard form of Newton's equation for gravity as:

$$F_A = -F_B = \frac{G.M_A.M_B}{R^2}$$

where

$$G = \frac{\phi}{\pi.\Theta^2}$$

Can the force of gravity be infinite for an infinitely massive body?

In deriving the above equation, a number of assumptions were made, which although valid when considering the force of gravity between the sun and the planets, are certainly not valid in all cases.

In the past three hundred years, since Newton's law of gravity was proposed, some people have tended to latch on to the equation itself, rather than the meaning of the equation, and have made a number of erroneous assumptions regarding the equation which have, at times, lead to absurd results. I would like to deal with a couple of the more erroneous assumptions and try to explain why these assumptions are incorrect.

If we consider Newton's equation for gravity, $F = \dfrac{G.M_A.M_B}{R^2}$, two different sources of error are possible.

The first is that the equation shows that as either of the masses increases towards infinity, the force between the two bodies would increase to infinity. The second is that as the distance between the two bodies decreases to zero, the force of gravity between the two bodies would increase to infinity. These two possible sources of error have actually been combined in some theories. By merely looking at the equation, one could easily start supposing that if one continually increases the mass of a body, the gravitational attraction of the body upon all its component parts will grow so large that it will compress the body, thereby reducing the value of R and hence increasing the force of gravity even more. By judicious use of the equation, one can arrive at all sorts of theories of black holes etc. where situations could arise whereby nothing could ever escape from a black hole once it is formed.

If one thinks of what actually causes gravity, one will soon realize that this is incorrect. For gravity is not actually a property of matter itself but is rather as a result of an imbalance in the aether between two pieces of matter as a result of the shielding effect of each of them upon the other. This shielding effect is in fact proportional to the cross sectional areas of both bodies rather than on their masses.

I earlier justified the use of the formula $\Theta = \dfrac{M_A}{X_A}$ where Θ is a constant throughout a large portion of the universe. In all of the discussions earlier, I have always been very careful to point out one of the most important assumptions used in deriving the equation for gravity. That is that the bodies that we are dealing with appear virtually transparent to the aether. This is a very important requirement, for only if this requirement is met can we assume that by doubling the number of particles, will the visible cross sectional area be doubled.

If we go back to my discussion of marbles falling on the floor, I stressed that the floor must be large enough so that the marbles can spread out without interfering with each other. Let us suppose that this was not the case. Suppose that we had so many marbles that instead of spreading out widely on the floor, the marbles actually ended up piling on top of each other to a depth of several feet in the room. In this case we would not be able to see all the marbles. In fact, if we looked down from the ceiling, the cross sectional area of the marbles that we could see would be the same as the area of the floor. Under such circumstances, the cross sectional area visible would no longer have any relationship to the mass of the marbles.

In the same way, if we were to increase the mass of a body continuously and reduce it's size, as might be the case in a black hole, we would start getting to the situation where the particles of matter in the body would get so crowded together that the chances of one particle obscuring another would no longer be negligible. Under such circumstances, one would have to revert to the more accurate equation which was used to derive the well known equation for gravity, namely:

$$F_A = F_B = \frac{\phi . X_A . X_B}{\pi . R^2}$$

where X_A and X_B are the actual cross sectional areas projected by the bodies upon each other.

Let us suppose that body B is a normal body that we are familiar with, such as the Earth. Lets suppose that we could somehow continually increase the amount of matter in body A and that body A were to gradually be compressed due to this enormous amount of concentrated matter. What would the effect of this be on body A and on body B.

Now remember that gravity is caused by the aether, not by the amount of matter in body A. Also remember that the effect of gravity is actually caused because of the difference in force of the aether on two sides of a body. When the particles in a body are far enough apart so that they never interfere with each other, then this force is proportional to the mass of the body.

Imagine that matter were continually being added to body A. As long as the particles in A were far enough apart that they did not interfere with the view that B had of each particle in A, the force of gravity would be according to Newton's equation for gravity.

But once the situation were reached where the particles in body A were so closely packed together that some of them started interfering with the view that B had of others, then one could no longer assume that the equation $X_A = \dfrac{M_A}{\Theta}$ is valid. Once this equation no longer holds, then Newton's equation for gravity will no longer hold. As an analogy with the marbles on the floor, once the floor was totally covered with one layer of marbles, further marbles would then fall on top forming a second layer which would gradually blot out the spaces between the first layer of marbles. Once two layers of marbles were present on the floor, there would no longer be any line of sight paths through the marbles. If we then added a third or more layers of marbles, the view from the ceiling would essentially be the same. A lot of marbles totally covering the floor so that no part of the floor could be seen.

So, by continually increasing the mass and density of body A, one would start getting to the stage where some of the particles in body A started blocking out the view from body B of other particles in body A. Once this starts happening, adding further matter to body A will no longer increase the projected cross sectional area of A, namely X_A, in proportion to the additional matter added. Eventually, once the density of matter in body A reached the point where there were effectively at least two layers of particles blocking the view of a piece of aether trying to move through body A, then effectively body A would have reached the point where all aether reaching body A would be reflected by body A. If one added further matter to body A without effectively increasing the dimensions of body A, there would be no further increase in the force of gravity from body A on body B despite the increase in mass of body A.

For in the end there is a very definite maximum to the force of gravity that can act upon body B. It is determined by the aether itself. One of our initial steps in deriving the equation for gravity was to let ϕ be the total pressure of matter in the aether on a flat surface. As discussed this value of ϕ is probably a constant throughout a significant region in the universe.

If we consider body B, of mass M_B and cross sectional area X_B, then it is clear that there is a very definite maximum force of gravity that could push upon body B. That maximum force would be if some other body was so close and massive that it completely prevented all matter in the aether from reaching body B from one side. In this case, the maximum total force of gravity on body B would be:

$$F_{B\_MAX} = \phi.X_B$$

If body B were a normal body, then

$$F_{B\_MAX} = \phi.\frac{M_B}{\Theta}$$

So as we keep on adding mass to body A, always keeping the distance R the same, our equation for the force of gravity upon body B resulting from the presence of body A would gradually shift from

$$F_B = \frac{\phi.X_A.X_B}{\pi.R^2} = \frac{\phi.M_A.M_B}{\pi.\Theta^2 R^2}$$

to become

$$F_{B\_max} = \phi.X_B = \frac{\phi.M_B}{\Theta}$$

After this point had been reached, increasing the mass of body A any further could not have any further effect upon body B, nor would moving body A closer to B.

Now let us look at how body A is affected.

All of the equations that have been derived since Newton's time for the center of gravity of a body implicitly assume that all particles within a body would be "visible" to an exterior body.

If body A becomes so dense that it effectively prevents all particles of the aether from moving through body A then a very interesting phenomenon will occur inside body A. Consider piece a matter, C, on the surface of body A. Since body A is so large that it prevents any of the aether from passing through it, it effective shields the piece of matter on the surface of body A from half the aether that would otherwise have passed by it.

Hence the force of gravity acting upon body C will, in the same way as for body B, be given by

$$F_C = \phi . X_C$$

where X_C is the cross sectional are of the piece of matter C.

Now the interesting thing is that if we keep on adding matter to body A, from the other side of A to where C is on the surface of A, the force of gravity acting upon body C will remain the same. In the same way it may be shown that, once body A has reached this critical density, at all points on the surface of body A the force of gravity will remain the same irrespective of how much mass is added to body A. And if body A were somehow to be compressed because of it's enormous mass, the force of gravity acting upon the piece of matter C on the surface of body A would remain the same irrespective of the radius of body A and despite the fact that the surface of body A would have moved closer to the center of body A, thereby reducing R.

Now think about a piece of matter, D, in the interior of body A. What force of gravity and what pressure would body D be experiencing? If one thinks about it, one will soon realize that if body A is so dense that it prevents all of the aether from passing through it, then there will be a certain "crust" at the surface of body A which reflects all of the aether. Once one gets below the level of this crust, no particles of matter in the aether will be able to penetrate. Hence below the level of this "crust" there will be no aether, just dense matter. Hence if the piece of matter D is below this "crust" it will actually experience no force of gravity whatsoever. Neither will any of the other pieces of matter below this "crust". The only forces acting upon body D will be due to the pressure all other pieces of matter about it all being pushed down by the force of the aether upon the crust of body A.

This is quite unlike our everyday experiences. We are so used to thinking of gravity as being progressive that it is hard to try to imagine what it would be like when gravity is not progressive. For example, we are all know that the deeper in the ocean we go, the higher the water pressure rises. The pressure in water increases with depth due to the cumulative weight of all the water above it. This cumulative pressure is due to the fact that the aether reaching any piece of water at any depth is hardly affected by the presence of the water above it and the Earth below it and each piece of water weighs as much as any other piece of the same mass. Imagine the situation, when considering body D somewhere deep inside an enormously massive body A, where the pressure only increases as we go deeper into the crust. One we get below the crust, gravity no longer exists and the "weight" of all bodies below the crust will be zero.

The pressure acting on all bodies below the surface crust will thus remain a constant throughout the interior of the body below the crust. In fact the pressure will be equal to ϕ, the force exerted by the aether per unit of cross sectional area.

What this means is that once this pressure has been reached, the pressure cannot increase further. This means that adding further matter to body A will have no further effect, due to gravity, upon bodies in the interior of A or on the surface of A.

This means that once a body has reached a certain density, then after that as the body becomes larger there will be no further increase of pressure within the body. The question posed at the start of this section was "Can the force of gravity be infinite for an infinitely massive body?".

The answer simply is that the force of gravity cannot be infinite, either outside the massive body or within the massive body. It reaches a maximum once the body becomes so dense that it prevents the aether from moving through the body. In the same way, even if the mass of a body grows towards infinity, particles of matter within the dense "crust" of the body will essentially be weightless and the pressure within the body will reach a maximum. Hence the density within such a massive body will also reach a maximum.

The maximum density that matter can reach would be that of pure solid matter as discussed in proposition 5 point 2.

Thoughts about the derivation of the law of gravity.

In this section, Newton's law of gravity was derived from first principles using Newton's laws of motion and assuming my concept that the aether is in fact the cause for gravity.

Although the two formulas are the same, the concepts behind the formulas are quite different.

In fact, if one studies Newton's Principia one will find that Newton was far more accurate in his description of what gravity was than many of those that followed him. In particular, Newton never actually wrote down the equation derived above in his Principia but rather talked in terms of proportions and inverse proportions. This is, to me, an indication that Newton probably considered the possibility of gravity being caused in the manner which I suggest.

One word was, however, introduced in the translation of Newton's Principia when talking about gravity which, while understandable, was unfortunate and needs to be revised.

This word is the word "attract".

This word "attract" leads, in my mind at least, to the notion that two bodies are somehow pulling upon each other in the same way that one's hand pulls on a door knob or the way in which the ends of a spring in tension connected to two objects pull the two objects together.

If my ideas on the reason for gravity are proved to be correct, then I think that the word "attract" should always be replaced by the word "push".

For the Earth does not attract us towards it by some magical gravitational property of the matter in the Earth itself.

Instead, we are rather pushed towards the Earth by the aether as a result of the Earth shielding us from a certain portion of the aether which would have counterbalanced the pushing force towards the Earth.

The choice of words in science is very important in conveying ideas, particularly when the words used are simple words commonly understood by the average person. The incorrect use of common words can cause a great deal of confusion and should be avoided whenever possible, even though it might not seem particularly important to scientists that understand the underlying theories.

Think of the simple word "flat" that is used to describe a level surface such as water. I can still remember the time when I was young when there were people who still believed that the Earth was flat. I can remember when the "Flat Earth Society" was finally disbanded once photographs of the Earth taken from space became available.

My feeling, therefore, is that it is most important that words such as "attract" and "repel", which are really being used to talk about things being pushed together or pushed apart, should be abandoned in science as soon as possible. For these words can easily and mistakenly imply that bodies somehow possess magical properties which enable them to pull things towards themselves or push things away from themselves.

For just as it has been shown that gravity can be accounted for as a result of the aether, which consists of small particles of matter moving at very high speeds, pushing bodies together, so I am sure it will be possible to eventually explain all the properties of electricity, magnetism and light as a result of some type of action of matter moving in the aether.

In the meantime, it would be better to talk of electrons being pushed away from each other rather than repelling each other. Similarly, two protons are pushed away from each other by the aether rather than being repelled by each other. An in the same way, a proton and an electron are pushed towards each other by the aether rather than attracting each other.

Once this distinction is understood, then better minds than mine will be able to start understanding why the protons within a nucleus of an atom are not pushed away from each other by the aether.

Newton's third law of motion and gravity.

The other thing that is most important to stress, when considering two bodies pushed towards each other by gravity, is that the things doing the pushing on each of the bodies are in no way related to the bodies themselves. The fact that the forces of gravity between two bodies is equal is due to the fact that if two bodies are stationary in the aether, then the laws of probability dictate that they will be pushed by equal and opposite forces towards each other.

Many people have incorrectly assumed that when considering gravity between two bodies, there is somehow a tension effect, such as a stretched spring, and that Newton's third law dictates that the force of gravity on one body must be exactly and equally balanced by the force of gravity upon the other body. This is totally incorrect and is an invalid application of Newton's third law.

I discussed this to some extent in Section 1 where I calculated an approximate speed of gravity based upon the way that the Earth and Moon are moving apart.

Now that I have presented my ideas on the cause of gravity, the reader will be in a better position to work out for himself when and under what conditions Newton's third law may be used. As was mentioned in Section 1, Newton's third law is applicable in situations where the time required for forces to propagate are negligible. Perhaps more importantly, Newton's third law is really only applicable when considering situations where bodies are connected, such as during impacts or by other forces which can be regarded as constants during the time of the experiment. Newton's third law is widely used in mechanics. This use is quite valid as the times for transmission of forces are negligible in comparison to the times taken for things of interest to happen. In addition, one assumes that in the mechanical members, such as pieces of steel etc., there are other constant forces binding the atoms in steel together in such a way that the atoms form a solid and rigid structure.

When one considers gravity acting between two bodies, we are dealing with something entirely different. We are actually talking of a large number of particles in the aether pushing a first body towards the second and a quite different and separate large number of particles in the aether pushing the second body towards the first. There is nothing common or connected about the two bodies other than that they are both in the aether and fairly close to each other so that they shield a measurable amount of aether from each other.

And there is nothing that requires that the force of gravity acting on the first body be equal and opposite to the force acting on the second. When this happens it is in fact due to the fact that the two bodies are moving so slowly with respect to each other and with respect to the aether that for the time taken for the aether to move between the bodies, the bodies have effectively not moved with respect to each other or the aether.

And just as these considerations hold with gravity, so I expect that we shall find that the same principles hold with respect to electricity, magnetism and light. I will deal with my thoughts on this in the third and fourth sections.

And this completes the purpose of this second section in which I postulated properties of energy and matter and gave an explanation of what the aether could be. From this I explained what the cause of gravity might be and then derived Newton's law of gravity from first principles. I then showed how Newton's law of gravity must be modified to account for situations not covered by Newton in his law of gravity.

Section 3 - Thoughts on the Universe

Thoughts on the size of the universe and the Big Bang Theory.

One or our current theories, The Big Bang Theory, postulates that our entire universe was once compressed into a tiny ball. At some instant of time this ball exploded and expanded massively into empty space to form our current universe. It is thought that this happened some 20 billion or so years ago.

I think that we need to re-examine this whole concept, particularly if my theory of gravity is correct. For if my theory is correct, there are a number of problems in the Big Bang theory that need addressing.

Firstly, what caused the entire universe to be compressed into such a small dense ball in the first place? As explained in Section 2, it could not have been due to the force of gravity. For as I discussed in Section 2, gravity has a finite force and cannot be infinite. And even if gravity were powerful enough to compress the entire universe into the ball, if the entire universe were compressed into the ball then would the aether also have been compressed into the ball as well? For the aether is an important part of the universe and could well contain most of the kinetic energy in the universe.

If the aether were compressed into the ball, then a number of problems arise. Firstly, there would have been nothing to cause the gravity required to keep the ball compressed. And what could possibly have compressed the aether into the ball? So what could have held the entire universe in such a compressed ball before the time that it started expanding. And even if somehow something had held all the matter in the universe in this ball, what caused it to suddenly let go. It would have to have been some external action for if it were not why would the universe have not remained in the ball forever? And how could it have been an external action if the entire universe were in the ball? Are we to imagine that there are other universes as well? And even if there were other universes, how did our universe initially end up in such a small dense ball. And when it did expand, what happened to the other universe that was the cause of the expansion?

Furthermore, as suggested in proposition 1, energy is matter in motion. Assuming that the entire universe, together with the aether, could have been compressed into this ball, then all the matter in the universe compressed within the little ball would have presumably have been limited to moving within this little ball. But if this had been the case, it is hard to imagine how matter could have been moving in this little ball to equal the kinetic energy currently present in the aether in the universe. For energy is not a property of matter and matter is not a property of energy. Energy is matter in motion. It is the movement of matter. And if all the matter in the universe had been compressed into a tiny ball, there could not have been much movement involved. For if all the aether was constrained in a tiny ball, I cannot visualize how the aether could now be moving at the speeds that it is today. So where did the energy come from to cause the universe to explode or to provide all the kinetic energy which is present in the universe today?

If the aether were not compressed in the ball, then a different set of problems arise. For in this case we would have the strange situation where most of the matter in the universe was within the ball but none of the energy could have been. Assuming that the aether had somehow held the ball together in the first place, what would have caused it to ever let the ball expand. For if the ball contained no energy, what would have caused the ball to expand at all and to fill our entire known universe? So one is left to conclude that the aether would have to supply the energy to cause the ball to expand. But if this is the case, how did the aether get all the matter into the ball in the first place? Did the aether perhaps change? And if so, why? And if the aether was outside the ball and for some reason caused the ball to expand, why did it expand all at once in a massive bang? Surely in that case the aether would have eroded the ball and gradually worn it away in much the same way that wind and rain erode mountains. If that were the case, I would expect matter to be leaving the ball in more of a steady state where matter were moving away at a constant rate rather than the situation were the further away matter is, the faster it appears to be moving away.

Whether or not the aether was compressed as part of the ball, current measurements in all directions suggest that we are actually very close to the exact center of the expanding universe. For measurements in all directions suggest that matter at the same distance form us in all directions is moving away from us at similar rates and that this rate increases in proportion to the distance from us. The probability that we should have so fortuitously ended up at the exact center of the universe seems to me little short of absurd. To me, this whole concept sounds too good to be true and reminds me of theories in the middle ages that assumed that the Earth was the center of the universe and that everything moved around the Earth.

As we manage to peer further and further into the universe, the time taken for light to reach us from those furthest distances has taken tens of billions of years to reach us. To my mind, at least, the galaxies that we are viewing at those furthest distances that we can see are remarkably similar to those of today which are very close to us. How did these galaxies form so quickly? And furthermore, what mechanism could possible have caused matter to move from the tiny ball to the furthest reaches of our universe, to get there so quickly that light that then left from those places is only now reaching us. For if light from the matter at the edge of our expanding universe is only now reaching us, then if the mechanism that moved the matter from a position almost contiguous to us in the "pre-bang ball" to a position such that light that left shortly after the big bang from those farthest edges is only reaching us now, then the mechanism must have moved matter at a speed considerably faster than light itself. And supposing that such a mechanism existed, are we to suppose that after such a tremendous explosion and rapid movement of matter, the aether would immediately be in the equilibrium required to ensure that light would suddenly start passing through the entire universe at a constant speed?

Perhaps we need to re-examine the reason why the big bang theory was initially proposed.

From what I can gather, there have been two underlying reasons for the Big Bang theory.

The first is that it has been observed that as we examine stars which are further away from us, the wave length of light lengthens in proportion to the distance that the stars are from us. This has been clearly determined by experiment and cannot be disputed. For certain atoms, such as hydrogen, helium etc. emit radiation at very specific wave lengths in the spectrum. The spectral patterns of these atoms can be recognized in light that we see from our own sun as well as light that reaches us from very distant stars. The spectral patterns of these atoms in distant stars is quite clearly recognizable but have very definitely shifted towards the red side of the spectrum. This shift has been called the Red Shift.

Now we know from experiment that when bodies are emitting light and are moving away from us, the wave length of the light lengthens in proportion to the speed at which the bodies move away from us.

The assumption was then made, in view of the fact that the wave lengths of light reaching us from the stars is shifted towards the red side of the spectrum, that all stars are therefore moving away from us and that the further away stars are from us, the faster they are moving away from us. Once this assumption is made, the only way that this can be explained is to assume that at some time everything in our universe was squashed into a small ball and then exploded. This would, if we were near the very center of the explosion, explain why everything appears to be moving away from us with a speed that depends upon the distance that stars are from us.

The second reason for the Big Bang Theory was the question of why is the sky black at night. The logic behind this question is interesting. Suppose that one were to assume that the universe was infinite and filled uniformly with stars. Obviously, the further away a star is the less light we will receive from the star. But it can be proved mathematically that if one were to integrate, or add up, all the little bits of light that we would receive from each star in an infinitely large universe then all the stars together would contribute far more light from all directions than our own sun does. Hence, according to the results supplied by this mathematical integration, it would be impossible for the sky at night to be black. But the sky at night is black. This clearly indicates therefore, in the minds of some, that the universe cannot be infinite. This is a very important requirement for the Big Bang Theory. For if it were possible for the universe to be infinitely large, the Big Bang Theory would be nonsense. For, by the definition of infinity, the initial ball in which the Big Bang theory assumes the entire universe was once compressed would itself have had to have been infinitely large. And if the universe is now infinitely large and was once compressed into a finite size ball and then expanded into an infinitely large universe, then matter at the infinite edges of the universe would have had to have moved infinitely quickly to get there. I will leave it to the reader to think about some of the other absurd contradictions would result if one were to assume that the universe was infinite and that the Big Bang occurred.

The fundamental question to be asked is whether it is logical to assume that, because the wave length of light changes very slowly depending upon the distance traveled, this must be due to everything in the universe moving away from us.

I do not think that it is.

For we know very little about light other than that which has been measured upon the Earth. We have various theories for light but in the end we do not yet understand what causes electrical charges or magnetism. Maxwell has clearly demonstrated that light is an electromagnetic phenomena connected with the aether. Gravity clearly affects light. It has been observed that the path of light can be bent when passing by large objects, such as when it passes close to the sun. The important thing to remember that it is not the sun which is attracting, or bending, light but rather the aether which is pushing the light towards the sun.

So an imbalance in the equilibrium in the aether, such as is caused by the presence of a massive body, can affect the motion of light. Given that light is travelling through the aether, I feel that it would be unrealistic to assume that it could travel for millions and billions of years through the aether without in some way being affected by the aether. For there could be various processes involved that we have not yet imagined which could explain why the wave length of light increases slowly as it passes through the aether. For to assume that light is not affected in any way by the aether as it travels through the aether is to assume that transmission occurs with perfect efficiency with no allowances made for any resistance or loss of energy for any reason whatsoever.

If we use standard equations and concepts of energy and conservation of energy that we have been accustomed to, the Moon should not be able to move away from the Earth. But it does. And I have been able to explain this due to the fact that there is a slight but cumulative amount of transmission delay between where rotating bodies actually are and where they appear to be due to the speed at which gravity moves.

Is it not possible that there could be an analogous small amount of transmission delay involved in the propagation of light, due to the finite speed of gravity and the aether, which is cumulative and makes itself manifest as light travels through the aether over a period of millions and billions of years? Is it not possible that this cumulative transmission delay can result in light loosing a very small amount of it's energy as it travels throughout the aether?

My feeling is that we must keep our minds open to the possibility that light actually does loose a very small amount of energy as it moves through the aether resulting in the "Red Shift". I will discuss, towards the end of this section, where I think this lost energy might be going to.

Suppose that the red shift in light is indeed caused as a result of some interaction between light and the aether and is proportional to the distance traveled. If the wave length of light does increase as a function of distance traveled, then this effectively means that light gradually looses energy as it passes through the aether. If that is the case, then there is a maximum distance which a ray of light can travel before it's wavelength increases to the point where the wave length becomes so long that it can no longer be detected. There would, therefore, be a very finite although extremely large limit to the distance to which a ray of light could travel through the aether.

If there is a very definite limit to the distance which light can travel through the aether, then this needs to be taken into account when trying to integrate the light that could be reaching us from an infinitely large universe. For if there is a maximum limit to the distance through which light can travel, then the assumptions made in the mathematical model that was used to integrate the light reaching us from an infinitely large universe would not be correct. For light could not reach us from most of an infinitely large universe as it would exceed the maximum distance that light can travel. I will leave it to the reader to prove that if the universe were infinitely large and if there is a finite distance to which light can travel, then sky at night will be black except for a small but constant amount of background radiation of very long wave length light which will be the same in all directions. I can remember when scientists accidentally discovered this background radiation as they were attempting to calibrate their instruments.

It is quite possible, therefore, that the universe is not expanding in all directions but could in fact be in a state of relative calm and equilibrium where matter is just drifting around lazily and randomly. In that case we would not be at the center of an expanding universe. Rather we would be at the center of the little sphere in the universe that is visible to us. This could in fact be a very little sphere in very much larger universe that we cannot and will never see.

There is insufficient experimental evidence at the moment to suggest whether the Big Bang theory is correct or whether the wave length of light does in fact increase with distance traveled, irrespective of whether the source of the light is moving away from us or not.

Until more evidence is available, I personally tend to favor my feeling that the Big Bang did not occur and that the universe is not expanding in all directions. I see no reason why one cannot imagine that the universe is infinitely large.

Thoughts on the age of the universe.

Because of the Big Bang theory, we have tended to think of the age of our universe as being determined by the time that it takes light to travel from the furthest distances that we can see.

If the Big Bang did not happen, then the age of the universe could be immeasurably greater than we have imagined. In Newton's time people thought that the universe might be ten or maybe even twenty thousand years old. Some people now think that it could be 20 thousand million years old but this number seems to be being continually revised upwards. My feeling is that these current estimates of the age of the universe could be nearly as incorrect as those in Newton's time.

Charles Darwin, in the nineteenth century, was perhaps the first person to truly understand how old our universe could be. I remember reading one passage of his where Darwin mentioned that the best history that we have of our Earth is that which is recorded in our genes over immeasurably long periods of time. As scientists are slowly starting to be able to read what is recorded in our genes, I am sure that Darwin will be proved correct. Perhaps it would be a good idea at present to try to preserve as many species as possible so that when we are able to understand what our genes and those of other species have recorded, we will be able to get a more complete history of the Earth.

My personal feeling is that in the same way that we will be able to get a history of the Earth and life on Earth from what is recorded in our genes, so we will be able to get a history of the universe by examining what is recorded in the elements in the universe and in large bodies such as galaxies. For the elements are composed of all sorts of isotopes with different half lives, some of which last for billions of years. And galaxies have certainly lasted for a very long time, perhaps far longer that we can imagine.

By examining how the distribution of isotopes of elements vary in particles of matter in deep space, far from our sun and galaxy, we may then be able to start getting some idea of when various pieces of matter found throughout the universe were formed. And by examining the shapes of galaxies and of matter far from galaxies that might have been part of those galaxies at one time, we will be able to form a better idea of the age of galaxies. The real problem, of course, will be in getting samples to measure and being sure of their origin.

While it may never be possible to determine the age of the universe, there may be a better and simpler method to determine the life spans of galaxies and stars if one assumes that the universe is in equilibrium with stars and galaxies being continually formed and that all stars and galaxies will eventually die.

Imagine that a butterfly, that might live for only a few weeks, wanted to try to find out how long humans lived for. Suppose that this butterfly had some mechanical means that would enable it to look at as many humans as it wished but that it had no ability to understand human speech or writing. What strategy could the butterfly use, in the few short weeks of it's life, to try to determine roughly the age that humans live to by simply looking at them?

The best strategy that I can think of would be for the butterfly to try to measure what fraction of humans are born each day and what fraction die each day. If it did this at present, it would find that more humans are being born each day than die. This would indicate to the butterfly that the human population was not in equilibrium but was increasing. Suppose the butterfly found out that on average, one out of every 15,000 humans all around the world died each day. On that basis alone, the butterfly could, as a very rough first approximation, guess that humans lived on average about 15,000 days or roughly 40 years. By more observation, and taking into account the difference between the rate at which humans are being born and dying and also assuming that all humans eventually die, the butterfly could make a better approximation for the life span of humans.

Imagine that we wish to measure the age of stars and galaxies and confirm that galaxies live for far longer than stars. We are in a similar, but worse, position to the butterfly for our lifetimes are to the lifetime of a star or galaxy far less than that of the butterfly is to us. One of the problems that we have is that it may well take a star hundreds of our lifetimes just to die and a galaxy far longer than that to die. On the other hand, however, we do have the ability to observe an enormous number of stars and galaxies in the universe. We also have the ability to record this information so that our descendents can see what conditions in the universe were like when we were alive.

If we can determine, by careful observation over a long period of time, what fraction of stars and galaxies are born during that time and what fraction of stars and galaxies die in the same period, then this would give us perhaps a more accurate idea of the age of stars and galaxies than we have at present. For our present methods of determining ages are based on guesstimates of the rate of decay of isotopes of elements on samples that we can measure and on current theories of stars and galaxies. Our knowledge in this respect is really limited to that which we have gained in just a few hundred years. To extrapolate that to billions of years is in some ways like having two points an inch apart on a piece of graph paper and then extrapolating the line to a distance of thousands of miles. More than a little bit of error will be introduced. Especially if the line is curved and we think that it is straight.

By studying the rate at which stars and galaxies are born and die, we will also be able to gain some experimental evidence as to whether the universe is in equilibrium or not. For if the universe were in equilibrium one would then expect that galaxies and stars die at the same rate that they are formed. Not only that, but we would expect that galaxies and stars are born and dying at the same rate in different regions of the universe.

Thoughts on galaxies.

The order that we see in galaxies of today suggests to me that they must be far older than we currently suspect. For if we take galaxies similar to our own, they have remarkably uniform and similar shapes. There tends to be a central core in the middle of the galaxy and a thin disk of material stretching out some 80,000 to 100,000 light years in diameter around this core. In addition, galaxies often have a spiral form where the disk of material actually consists of two or more spiral arms winding away from the center of the galaxy.

If we think of our own milky way galaxy, I have read estimates that it takes our sun about 240 million years to go around the galaxy once. I am personally not sure about that figure. Nevertheless, if one takes current assumptions that the universe is about 20 billion years old and our sun revolves about the galaxy every 240 million years, our own galaxy could not have even rotated 100 times about the axis of the galaxy. When one considers the forces acting on stars rotating about the galaxy, they are extremely weak due to the vast distances involved. Given the uniformity and sizes of galaxies, I feel that it is strange that they could all have arrived at such uniform shapes in less than 100 rotations.

Recent evidence from the Hubble telescope suggests that galaxies which are present at the furthest distances that we can see are in pretty much the same form that galaxies exist very near to us. Some scientists are puzzled that galaxies could have formed so early in the history of our universe.

I am not at all puzzled and rather feel that we are starting to gather experimental evidence that the Big Bang did not occur and that what we are seeing are galaxies at the furthest distance of what is visible to us.

I wonder, if we could somehow be immediately moved to one of those furthest galaxies, what we would see when we looked at own galaxy through telescopes. My feeling is that could well see our own milky way galaxy as it was twenty billion years ago and that it would look very similar to the way that it does today.

Over the years I have wondered why galaxies have the shapes that they do. In particular, why they are flat discs, often with curved arms.

Try to imagine what makes a galaxy flat. I have seen pictures of galaxies passing through each other with very little change to either of them. If a galaxy is not even able to materially affect the movement in another galaxy through which it is moving, what could have caused either galaxy to assume the flat disc shapes that they have?

Think of a galaxy eighty thousand light years across. The matter at the edges of the galaxy all lies almost on the same plane. What could cause this? If one tries to imagine that the matter was initially in a large sphere, what would have caused it to collapse into a disc in an almost perfect plane and how long would this process have taken? Perhaps one could imagine some strange set of circumstances that might explain why one particular galaxy had a flat disk shape. But nearly all galaxies are of similar shapes. So we are not dealing with some unusual combination of circumstances but rather something which is remarkably consistent across galaxies.

I wondered about this over the years. A possible explanation occurred to me one evening when my wife had taken me and our children to see a dinner and play on my birthday in about 1998. I must admit that I was not popular with my wife when she asked me, during the play, what I was thinking about and I mentioned that I had an idea about galaxies.

What occurred to me that evening was that a galaxy might have, at its center, a very large massive core. This core might then be causing the aether to push matter towards it from all directions. If one looks at pictures of galaxy M104, the Sombrero Galaxy, taken by the Hubble Telescope one will get an idea of what I mean. Please look at the photo marked Section 3 – Photo 1, taken by Nasa scientists using the Hubble telescope, shown below.

Section 3 – Photo 1 - The Sombrero Galaxy

There is a very bright central core and an "atmosphere" visible on both sides of the planar disc that that is almost half the diameter of the galaxy. This atmosphere is glowing somewhat but is very nebulous as one can see through it to the far edge of the galaxy at the back.

Imagine that this matter in the "atmosphere" of the galaxy is gradually getting pushed by gravity towards the central core from all directions. As the matter gets closer to the central core it will be accelerated by gravity pushing it towards the central core. As it gets closer to the central core, the matter would then reflect light from the central core as well as the flat disk. This could explain the glowing "atmosphere" seen around the core in the sombrero galaxy.

What we could have then is a situation where matter is continually approaching this central core from all directions, accelerating continually due to gravity as the matter approaches the core.

A point will be reached where the matter eventually arrives at the central core and adds to the matter already in the central core. The core will therefore be a massive body of matter where matter is continually being added to it. Depending upon the source of matter arriving at the central core, the core itself may or may not be spinning rapidly.

Some scientists have theorized that at the center of a galaxy there might be a massive black hole. At present there is insufficient experimental data to decide this question one way or the other. What is more important, to my mind, is what happens to all this matter that is continually being added to the core of a galaxy from all directions. One scenario is that the black hole simply sucks matter into it which can never escape again.

I disagree with this concept for many reasons. Firstly, if black holes simply grew and grew, one would expect galaxies to simply grow and grow. In such a situation, one would expect a wide variation in the sizes of galaxies. But this is not the case in practice.

Secondly, if my concept of gravity is correct, there is a very definite maximum "strength" that gravity can have. As discussed in Section 2, once a body reaches a certain density, there is effectively no gravity whatsoever within the body. If one keeps on adding more and more matter to such a body, which is spinning, what will happen? In the end, matter will be spun off in pretty much the same way that sugar is spun off in a cotton candy (candy floss) machine.

So, irrespective of whether or not black holes exist at the centers of galaxies, if matter is continually being added to the core of a galaxy, a point will be reached where there is so much angular momentum present that gravity will simply not be able to keep everything together.

What could we then expect to happen? My feeling is that one must expect that in one way or another, once the central core has reached a certain size and speed of rotation, matter will be ejected from the central core. If this is the case, then what would be the most probable direction for matter to be ejected?

If one thinks about a spinning mass of matter, the most probable point for matter to be ejected will be where the angular momentum is the greatest. This will be on the plane through the center of the core perpendicular to the axis of rotation. Hence one can expect that once matter has finally arrived at the central core of the galaxy it will eventually, irrespective of what fusion or other reactions it takes part in, be ejected from the central core in the direction of the plane through the center of the core and perpendicular to the axis of rotation. Not all of the matter will be ejected in the above mentioned plane. But the most probable direction would be in the plane with the probability dropping of sharply as the angle away from the plane is increased. One might, therefore, have a slight range of angles on each side of the plane from which matter is ejected, evenly distributed on both sides of the plane. As this matter moves out from the core of the galaxy it is quite possible that the matter on each side of the plane would attract the matter on the opposite side of the plane thereby causing it to gradually approach the plane itself.

And just as there are limiting factors that determine the size of hail stones or snow flakes, so in the case of galaxies there are limiting factors that determine the size of galaxies.

In the case of galaxies, the limiting factor may well be related to the maximum force of gravity possible. This determines, for any speed of rotation in the central core, the maximum amount of matter that can be present in the central core before some of it starts getting thrown out of the central core.

This would explain why galaxies have a very definite range of sizes.

This would also explain why galaxies have a flat disk like structure. For the matter in the disk of a galaxy consists predominately of matter that has been ejected from the central core of the galaxy and is slowly moving away from the central core.

The Sombrero Galaxy, because of the fact that plane of ejection lies almost along our line of sight, gives us a very clear picture of what is going on as far as matter moving towards the core of a galaxy. The halo in all directions around the galaxy is almost elliptical in shape and extends slightly past the dark outside band of matter at the edge of the shining disk. In addition, the halo is clearly very nebulous as one can see through it to the far edge of the galaxy. Both of these things indicate to me that this halo is made up of very nebulous matter that is reflecting light originating both from the core of the galaxy as well as from stars in the disk itself. By nebulous matter, please do not think that I am thinking only of dust. For old suns, planets etc. that eons before left some other galaxy might eventually make their way back into the core of a galaxy. By nebulous, I simply mean that over a given volume of space, the overall density of matter in that volume is extremely small. The halo around a galaxy is in effect similar to the tail of a comet in that it consists of very nebulous matter that is reflecting light rather than producing light. The elliptical shape is due to the fact that as one gets further away from the core, the fraction of light being reflected from the disk increases in relation to that from the core until one gets to the edge of the galaxy where virtually all light reflected is light from the glowing disk.

To get an idea of what is happening to matter being ejected from galaxies, the best idea will be obtained from those galaxies where we can view them from a direction in the line of their axis of rotation.

Galaxy M77 is a good example that shows how matter is moving away from the central core of the galaxy. I have included a photograph of this galaxy taken from Wikipedia. This photograph was taken by Jyri Naraden and Raine Karjalainen and released into the public domain. There are two different examples in this galaxy that indicate what might be happening. Please examines the photograph shown as Section 3 - Photo 2.

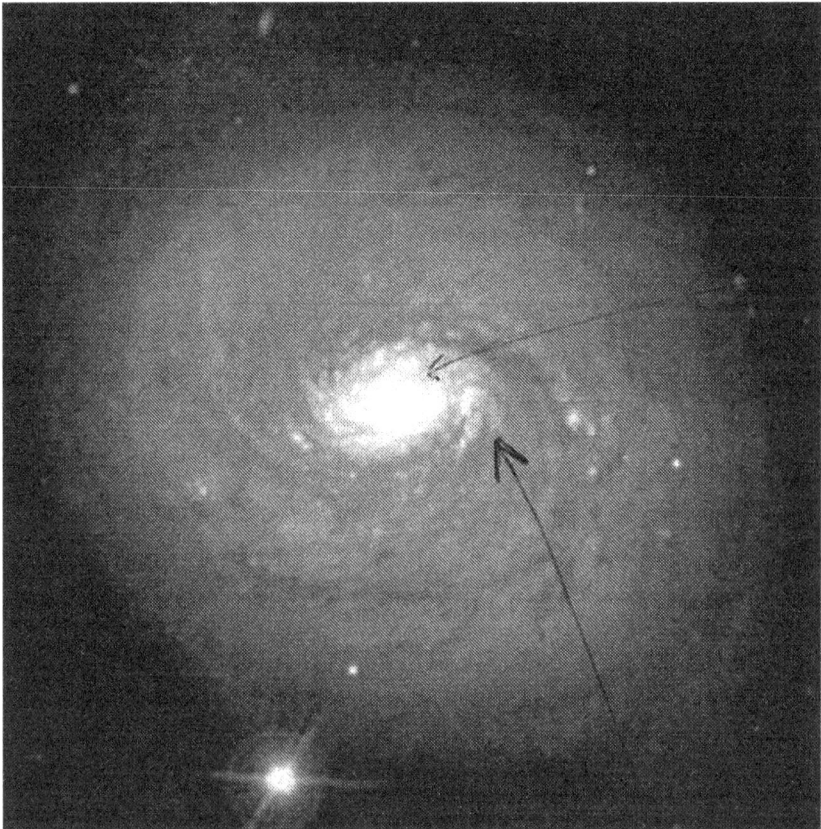

Section 3 - Photo 2 - Galaxy M77.

The central core of the galaxy is roughly elliptical in shape. At the left side of the ellipse, there is a jagged line looking almost like a sideways M, with some bright matter inside the M. If one examines this carefully, it is quite clear that matter is moving out of the gap. To the left of the M, about the height of the M away, there is a little piece of bright matter which seems to have moved away from a piece of the central core now at the eight o'clock position. It looks to me as if that piece of matter broke away when the edge of the central core, now at the eight o' clock position, was originally at the nine o'clock position.

On the right side of the central core, there is an even more interesting piece of bright matter that has broken away from the central core. If one looks at the picture, it appears to me as if the matter broke away from the gap in the central core currently at about the one o'clock position when it was at the three o'clock position. In both of these cases, matter has moved away from the central core as opposed to possibly moving towards the central core. Galaxy M77 appears to me to be rotating anti clockwise very slowly in relation to the speed at which matter is moving away from the central core. The bright piece of matter that has broken away from the central core seems to have broken away when the gap in the central core was between the three and four o'clock position and seems to have moved almost in a horizontal position from left to right as the central core rotated. If one looks further to the right, one will see another part of a spiral arm of the galaxy. If one traces the arm around the galaxy, it almost appears to me as is it was once attached to the central core when the ten o'clock position of the central core was at the three o' clock position. If this is indeed what happened, then that part of the arm has moved almost half way out of the galaxy during the time that it took the matter at the edge of the central core to rotate about half a revolution. If this is so, then matter in this galaxy is moving out of the visible region in just over one rotation of the central core.

One question to answer, if this is so, is whether it is possible to develop a mathematical model to describe how matter in the disks could be moving away as the evidence in galaxy M77 might indicate. This is a very important question than needs answering. The way that one uses mathematics often depends upon how one views a problem. One generally develops a mathematical model of a process in accordance with one's understanding of what is happening in a process. In the case of a galaxy there are a number of unknown variables that enter into the mathematical equations, such as the mass of the central core of the galaxy, the mass of matter in the disc itself, the age of the galaxy etc. The big thing to understand is that the best records that we have of any galaxy only go back a couple of hundred years. When one considers that it might take a galaxy over 250 million years to rotate once, we clearly do not have much in the way of historical data to see how the arms of any particular galaxy have moved in the past. The best that we can do is to make informed guesses.

One might easily assume that a galaxy was a fairly stable arrangement of matter where the matter in the disc simply rotated about the core forever. Once this assumption is made, one could then develop mathematical equations and determine what the mass of the core might be to try to account for the observed motion. On the other hand, if one assumes as I do that the discs of galaxies are planes in which matter is being continually ejected from a central core then one would have to develop a very different set of mathematical equations to model the galaxy. It may well be that the discs are rotating very much more slowly than previously imagined and that matter might be moving away from the core so quickly in relation to the speed that it is rotating about the core that in some galaxies the arms may not be rotating much at all.

For if matter is continually being ejected from the central core, then we are not dealing with a system in equilibrium in the same way as if we assumed the matter in the discs rotated in the same orbits forever. For what will be happening, on a continuing basis, is that matter is being ejected from the central core of the galaxy. This matter need not be at the escape velocity of the galaxy itself. All that is required is that it be slowly moving away from the central core. As additional matter is ejected from the core it will try to displace matter already in the disk. Where would this displaced matter in the disk go? It would not be displaced perpendicular to the plane of the disk for two reasons. Firstly, the plane of the matter in orbit about the core would have to change. Secondly, there will be a gravitational effect whereby material already further out in the disk will cause gravity to push material in the disk that is closer to the core towards the matter in the outside of the disc. So what will happen, therefore, is that as additional material is ejected from the core into the disk, this material will displace material already in the disk. The displaced material will be displaced in the direction of the plane and will move further out from the core in the plane of the disk. This will in turn displace material further out and so on. So although material in the core will initially be ejected with barely sufficient energy to leave the core of the galaxy, because of the continuous accumulation of material in the disk, there will be a gradual and steady movement of material from the core of the galaxy to the furthest edges of the disk.

Lets now consider how fast the material will move out from the core of the galaxy to the edge of the galaxy. If galaxies are very stable objects, which I suspect, then one should expect that there will be an equilibrium set up, particularly if the amount of matter reaching the core of the galaxy from all directions per unit of time remains fairly constant for billions of years. In this case, the amount of matter arriving at the core of the galaxy from all directions in space in a certain time should, provided the period of rotation of the core does not change, exactly equal the amount of matter leaving the core of the galaxy and entering the plane of the disk of the galaxy. In the same way, the amount of matter leaving the core of the galaxy should exactly equal the amount of matter moving out of the surface of a cylinder with the axis of the cylinder passing through the center of the core and perpendicular to the disk of the galaxy and of any radius. What this means is that if one considers any annulus in the disk, the amount of matter entering at the inner edge should equal the amount of matter leaving at the outer edge. As the radius of the annulus increases one of two things should happen, depending upon the original angular momentum with which matter left the core. Either the disk will become uniformly less dense as one moves further from the core or, alternatively, the matter in the disk may gradually condense into a number of "arms" which swing out from the core by centrifugal force and are held together by gravity caused by matter in the arms themselves. Nevertheless, if the galaxy is stable, the total matter ejected from the core per unit of time should equal the total amount of matter moving along all the arms.

Galaxy M51 is a very good example that illustrates not only how matter moves out from the center of the galaxy but also how a neighboring galaxy has in fact accelerated an outer arm of the galaxy such that it is moving faster than the inner core from which it escaped. I have included a photo, marked Section 3 – Photo 3, taken by NASA scientists using the Hubble Space telescope.

If one looks at the picture of the M51 galaxy, there is a very dense central area which appears solid on the photograph. The core itself would be at the center of this dense central area and could be very much smaller than this central area, perhaps even invisible. For the purpose of discussion, imagine the diameter of the "solid" central area as being the approximate distance across the visible solid central area before any gaps start appearing. At the end of the solid central area a dark jagged gap appears, at about the eight o'clock position, which then follows a spiral path away going clockwise for about 135 degrees and at a distance from the center of the core from about the diameter of the central core to about one and a half to two times the diameter of the central area. If one examines this gap closely, one will see that the width of the gap is very uniform even though the gap itself is rather jagged. Towards the clockwise end of the path of the gap, between the two and three o'clock positions, there is a jagged little piece almost the shape of a tooth on a saw. There is no doubt in my mind, when looking at that little saw tooth shaped gap, that the dense matter on the outer edge of the gap was once joined to the dense matter on the inside edge of the gap.

Section 3 – Photo 3 - Galaxy M51.

This dense matter on the outside edge of the gap has broken away from that on the inner edge of the gap. I can think of no other way to explain the existence of such a remarkable saw tooth shape with such a uniform gap. In fact, if one looks at the picture very carefully, one can actually see that the matter on the outside edge of the gap, at the saw tooth, actually seems to be rotated slightly anti-clockwise from the material on the inside edge of the gap.

If one looks at M51, from the center of the galaxy about 90 degrees anti-clockwise from the saw tooth gap, there is a large bright spiral arm of the galaxy further out from the core about two and a half times the diameter of the dense central area from the center of the galaxy. If one moves clockwise around this spiral arm one gets to a last bright piece of the arm. It almost looks to me as if this piece could have once been connected to the dense piece of the galaxy on the outer edge of the saw tooth at one time. If that is indeed the case, the whole outermost arm of the galaxy could have pulled, by the neighboring galaxy, from the position of the saw tooth.

In fact, if one really examines galaxy M51, one will see that the edge of the arm mentioned has been pulled up quite considerably from where it originally was by the neighboring galaxy. The reason for this is that the central core of galaxy M51 appears to me to be rotating anti-clockwise. The material at the outermost arm has actually been pulled anti-clockwise at a faster rate than the central core is rotating. This should give one some idea of how slowly the central core of galaxy M51 is rotating.

There are other galaxies, such as M58, M61, especially M63, M74 and M101 where observant readers will be able to see for themselves clear evidence that the matter in the discs of these galaxies is in fact moving away from the centers of the galaxies. For all of the above mentioned galaxies clearly show gaps of uniform widths but jagged shapes that clearly indicate that they were formed by matter that had been contiguous and has broken up and moved away in separate directions as it gets further away from the central core. Galaxy M61 has a very clear saw tooth shaped gap at the start of one of the arms in the galaxy. The central core is elliptical and the gap is about two times the distance of the major axis of the core from the center of the core and perhaps 15 degrees clockwise from the direction of the major axis.

Section 3 – Photo 4 - Galaxy M63

Galaxy M63 is a wonderful example of a galaxy where the effect of matter moving away from the central core can be clearly seen. I have included a photo, marked as Section 3 - Photo 4, taken from Wikipedia of Galaxy M63. There is very little in the way of spiral arms outside the core. The galaxy appears to me to perhaps be rotating very slowly clockwise which has resulted in the disk splitting up very uniformly into a large number of pieces with clearly defined gaps between the pieces, almost like a jig-saw puzzle. If one looks at the front of the disk one will see a large number of gaps of equal widths throughout the disk. There can be no doubt, when looking at the disk of M63, that the material has moved, as a single sheet, from the central core and has gradually broken up into a large number of interlocking pieces as the matter moves further from the core and has to, so as to speak, stretch out to cover the extra area.

My feeling is that whenever one sees saw tooth gaps between pieces it gives one a very good indication that the two parts separated by the gap were once joined together. For if one looks at Africa and South America, one gets the same saw tooth effect and one can readily imagine how South America and Africa were once joined together, particularly if one asks what could have resulted in such a similarity of shapes if they were not originally joined together.

It appears to me, therefore, that the flat disc shapes of galaxies are because the discs are planes in which matter is being ejected from the central cores of galaxies. And it may be possible, once this possibility is accepted, to look for the evidence of material which is no longer generating light or being lit by reflected light but which is still in the plane of the disk of material that has been ejected from the core of the galaxy. If one could find out how far these discs really extend into a dark region, it might be possible to get some idea of how old galaxies really could be.

If the planar discs of galaxies are in fact found to be planes where material is being ejected from the cores of galaxies, then I think that this provides very clear evidence that there is, in fact, a maximum intensity that gravity can have. This would be, in some ways, a justification for my theory that gravity is in fact caused by the aether and is not a property of matter in the way that has been previously thought.

When one looks at picture of the "Evil Eye" galaxy it is apparent to me that the core of that galaxy is ejecting material very rapidly in comparison to most galaxies. I have include a photo of it taken by the Hubble Telescope marked as Section 3 Photo 5.

Section 3 – Photo 5 - The Evil Eye Galaxy

I think that what must have happened is that by some means, such as the core of the galaxy having a close encounter or collision with the core of another galaxy or even the non visible disk plane of an old galaxy, or by the cores of two galaxies orbiting about each other very closely, matter is being fed from one galaxy into the core of the "Evil Eye" galaxy at a very much faster rate than matter is normally fed into the core of a galaxy. If one imagines that the core of the galaxy cannot increase, due to the finite limit of the strength of gravity, the core would probably not be able to grow much larger than that of a regular galaxy before reaching the point that it cannot accumulate any more matter and has to eject matter into the disk of the galaxy at the same rate as the matter is being received. In this case, therefore, the thing that is going to grow in size is not the core itself but the amount of material in the disc. The disc will become thicker and much denser. So dense in fact that we might not be able to even view stars shining in the disc itself. What we may even be witnessing is the death of a galaxy as it is swallowed up by another galaxy.

Thoughts on our future in the Milky Way Galaxy.

One analogy that comes to mind when trying to imagine how galaxies work is to think of the Antarctic ice shelf as the core of a galaxy. Water molecules are continually being added to the ice shelf by the atmosphere in an analogous way to which matter is continually being added to the core of a galaxy by gravity. Once a sufficient amount of water has built up in the ice shelf, it then slowly moves away towards the sea. Pieces of the ice shelf break off as icebergs and slowly drift around the Antarctic and eventually melt away as they move further and further away in an analogous way to which large chunks of matter break away from the central core of a galaxy and then drift away in the disc of ejection of the galaxy and gradually drift off into space and finally melt away. And in the same way that the time that water spent in the ice shelf of the Antarctic could be thousands of times longer than the time that it spent drifting away from the Antarctic once it broke away, so might the time that matter spends in the core of a galaxy be thousands of times longer than the time that it spends drifting away from the core of the galaxy once it breaks away from the central core.

If galaxies work as I have suggested, then it is quite possible that galaxies are very stable arrangements of matter that could have existed for far longer than we have hitherto imagined. If my concepts about galaxies are correct, then just about everything that we see in the arms of our milky way galaxy were once part of the central core of the milky way galaxy and have been ejected from it. This means that we are continually moving away from the central core of our galaxy. What future awaits us?

In the long run our sun will gradually move away from the center of the milky way galaxy and one day drift into those far reaches of the galaxy where the stars no longer shine.

If humanity is to survive at that time, our first strategy will be to look for hospitable suns and planets closer to the center of our milky way galaxy. For if our galaxy is stable and far older than hitherto supposed, a very stable long term solution presents itself, at least until such time as we are ever able to travel between galaxies. That solution, as our sun and subsequent suns continually move away from the center of the galaxy, is to continually move towards new suns which are closer to the center of the galaxy. By continually moving along the arms of the spiral towards the center, it might be possible to continue living within the galaxy long after our own sun has moved into the dim extremities past the visible edge of our galaxy.

And if other more advanced forms of life already exist in our milky way galaxy, their first strategy will surely be the same. And if one imagines that more advanced forms of life will have had longer to develop, we should expect the more advanced forms of life to be further from the center of our galaxy than we are. And if such advanced forms of life exist and ever need to and have the ability to move closer to the center of the galaxy, it would probably be a good idea not to advertise our position. For those persons interested in searching the stars for life, I would recommend looking for signs of life in the outer edges of our milky way galaxy rather than towards the center. For if we ever do get invaded by aliens, which I doubt, they will probably be arriving from the outer edges of our galaxy by moving along the spiral arm towards us.

But in the meantime, perhaps the first order of business should be to try to determine how old our galaxy really is so that we can work out how much time we really have.

We have to be very careful, when taking measurements to try to determine the ages of things, that we do not fall into the trap of assuming that measurements that we make apply to all regions of space.

To give an example. At the moment we have samples of material from the Earth and the Moon. If the Earth and Moon were originally one body, as I suspect, then it is quite possible that whatever giant impact caused the Earth and the Moon to separate could have well had an effect on all matter that we sample from the surface of the Earth and Moon. It is quite possible, when we start doing radioactive dating of samples, that we will find that many things appear to date to the same time period, namely when the Earth and Moon separated. We would make a major mistake in assuming that such readings in any way indicated the age of the solar system.

In a similar way, once we can get samples of material from outside our own solar system, we may once again find that much of it tends to date from a similar but even longer time period ago. And once again we could make a serious mistake if we were to assume that this somehow gave the age of the galaxy or the universe. For all the material in the area of the arm of the milky way that we are in probably left the central core of the milky way at about the same time. We could then merely be measuring the time when our section of the milky way left the central core of our galaxy rather than how old the milky way galaxy itself is.

If one really wants to get an idea of the age of the universe, the best material to try and sample would be that material which is arriving from deep space and headed towards the core of the milky way rather than material which has been ejected from the core of the milky way. For material entering the core of our galaxy could well have been ejected many times from the cores of other galaxies over the eons. The practical problem in doing this, of course, is that we are at present unable to get to the positions around our galaxy needed to collect the samples.

One interesting thing that should be considered is that if all the stars visible in our milky way galaxy were originally ejected from the core of the galaxy and are gradually moving towards the edge of the galaxy, then most of these visible stars would also be moving away from us. It is to be expected, therefore, that there would be a slight red shift of light from the stars in our own milky way galaxy due to the fact that these stars really are moving away from us. This red shift would depend on their positions within our galaxy with respect to us. Over and above that, there would be the second red shift due to the distance that the light has traveled. By measuring red shifts carefully and trying to distinguish what could be due to matter moving away from the core of our galaxy as opposed to that caused by distance traveled, we might be able to determine the speed at which matter is moving away from the core of our galaxy. This would give us some idea of how long it takes matter to leave the milky way galaxy and how long it would have been since the matter in our sun was ejected from the core of our galaxy.

From a longer term point of view, it would be helpful if we could get observation points on either side of the disk of our galaxy and away from the plane of the disk so that we could start examining material headed for the core of our galaxy. For in the long run, the future of our galaxy depends upon what enters the core and at what rate.

Thoughts on our Solar System.

When one considers our Solar System, the same interesting facts appear as appear in galaxies. All of the planets in the solar system lie very nearly on one plane which passes through the center of the sun and is perpendicular to the axis of rotation of the sun.

This raises an interesting possibility in my mind. Is it perhaps possible that all of the planets lie in the same plane because of the fact that they were all formed from matter originally ejected from the sun itself. For the sun itself may act in a similar way, on a much smaller scale, as a galaxy.

Is it possible that the sun could be attracting matter from all directions, which would be a small volume in the arm in the milky way that we are situated in. Could this matter be used for fuel and at some point ultimately be ejected from the sun. If it was ejected from the sun, such as by means of solar flares, the matter that has the most chance of escaping from the sun would be that which was ejected at the point where it's angular momentum would be the greatest. That direction would be on the plane through the center of the sun perpendicular to the axis of the sun, namely the plane where the planets are located. Is it possible that the solar wind could slowly, over a period of billions of years, gradually move this matter towards the furthest edges of our solar system and in doing so, gradually form the planets?

At present we do not yet have the technology to measure how quickly the planets are moving away from the sun. But just as the Moon is destined forever to move away from the Earth due to the finite speed of gravity, so are the planets destined to continually move away from the sun. Is it possible that all the matter in the planets around the sun and in the asteroids which lie in the plane of the planets past the orbit of Pluto, were once all ejected from the sun? If so, the sun must be far older that we currently think, but certainly not as old as the milky way galaxy itself.

I think that it would be most useful if we could launch satellites to travel at much greater speeds than they have hitherto done, to be able to get well beyond the orbit of Pluto and to try to get an idea of how far out our solar system really extends and whether there are any other as yet undiscovered massive planets out there, in the plane of the other planets, slowly orbiting about the sun and so far away from us that they reflect an indiscernible amount of light.

And it might be most useful to send satellites away from the sun in a direction perpendicular to the plane of the planets so that we could get a better view of our solar system. For it may well be that ring like structures, which are in the plane of the planets and invisible from Earth, surround the sun both very close to the sun and also far past the orbits of Pluto and the furthest known objects in the solar system.

One idea that I have had to try to get a glimpse of how far our solar system extends would be to launch two satellites and position one satellite on each side of the sun along the axis about which the sun and all the planets rotate. The two satellites should be positioned at least three billion miles, or about four and a half light hours, above the sun, one on each side of the sun along the axis about which the sun rotates. Very occasionally, one gets massive explosions in space which cause tremendous amounts of light to be released. I believe that one such event occurred recently where the light was so bright that it was possible to detect it being reflected from the surface of the Moon.

Suppose that the two satellites were set up to detect such rare explosions. We might be able to use these explosions in a similar way to a flash bulb on a camera, which lights up a subject for a very brief time sufficient to enable a photograph to be taken. Once the satellites detected such an event, they could locate the direction that the flash arrived from and it's intensity. One could then calculate, based upon the speed of light, the time at which any point in the plane of the planets about the sun would have received that light. Knowing the distance from the satellite to the sun, one could then calculate how long it would take that light to be reflected by anything which happened to be at that point in the plane to arrive back at the satellite. Each satellite would then have several hours to be able to position itself to take a series of photographs of the reflection of light coming back from various distances out in the plane of the planets. In addition, the satellites would be able to calculate the probable intensities of lights from different points in the plane and adjust the camera sensitivity accordingly. If we are lucky, we might be able to get a brief glimpse of giant rings of dust and debris extending tens of billions of miles from the sun and maybe, with extreme luck, even get a glimpse of other large planets well beyond the orbit of Pluto.

Thoughts on the sun itself.

Most people seem quite unaware of what a massive body the sun itself is. When I ask people what percent of the sun's energy reaches Earth, most of the people that I have asked say maybe one or two percent. A few have said half a percent but nobody has ever said less that one hundredth of a percent. I find this lack of a basic feel for our solar system staggering.

It is quite simple to work out a very good approximation for the fraction of energy from the sun that reaches the Earth. Let us suppose that the Earth moves in a perfectly circular orbit around the sun at a distance of 150,000,000 kms from the sun. If we think of a sphere with center at the center of the sun and radius 150,000,000 kms then all the energy radiated by the sun would pass through the surface of this sphere. Now imagine our little

Earth, at a distance of 150,000,000 kms from the sun. The fraction of the total energy of the sun that the Earth receives would simply be the cross sectional area of the Earth facing the sun divided by the area of the sphere with radius of 150,000,000 km. Now the cross sectional area of the Earth facing the sun is just that of a circle of radius 6,371 kms.

Hence, the fraction of the sun's energy reaching the Earth

$$= \frac{\text{Cross Sectional Area of Earth}}{\text{Surface area of sphere about the sun}} = \frac{\pi.6371^2}{4.\pi.150000000^2}$$

$$= \frac{6371^2}{4*150000000^2} = \frac{4.059*10^7}{9.0*.10^{16}} = 4.5*10^{-10}$$

This fraction is thus 4.5 ten billionths. The percentage of the sun's energy that reaches the Earth is thus only 45 billionths of one percent. Looking at it the other way around, the sun puts out 2.2 billion times as much energy as the entire Earth receives from the sun.

To give you some idea of what this means, in one second the sun puts out as much energy as our entire Earth receives from the sun in just over seventy years. In two years the sun puts out more energy than the entire Earth has received in the last 4.4 billion years.

Let us think about what sort of problem this would pose to future generations in space who might be trying to communicate by radio with Earth. Imagine a colony of adventurers who have traveled to our nearest star, some four light years or just over two million light minutes away. It takes light eight minutes to reach the Earth from the sun. So think of a triangle with a base of 8 units and a height of two million units. Now imagine that you are at our nearest star at the very apex of the triangle looking down at the base of the triangle which is eight units wide and two million units away. At the one side of the base of the triangle is our sun which is pouring out more energy each second that the entire

Earth receives in seventy years. And at the other side of the base of the triangle is our tiny little Earth with humans on it trying to send out radio waves with puny little radio transmitters where power is measured in kilowatts. Irrespective of where the Earth is with respect to the sun, the sun is just pouring out so much more energy that any radio waves that we attempt to send will literally be swamped by the energy radiated by the sun. And the further away a star is from our sun, the narrower will the angle between the sun and the Earth become and the more will radio waves from the Earth be swamped by the light from our sun.

In order to communicate with fellow humans at our nearest star, it would be necessary to set up a whole system of relay stations where the first purpose would be to try to increase the distance between the relay stations and the sun. In addition, if we are to stand any chance of our radio waves being measurable from the nearest star, we would have to develop a most sophisticated system of focusing all the energy being transmitted as narrowly as possible and aimed as accurately as possible at the nearest star.

And if aliens existed in our galaxy and were attempting to communicate with their brethren on other stars, they would face the same two problems. Firstly to try to prevent their own stars from interfering with their communications and secondly to aim and focus their communications as accurately as possible. In both cases, if one thinks about it, it is almost unimaginable that we would ever be able to detect radio waves from aliens even if they were as close as our closest star. For if the aliens made no attempt to shield their communications from their sun then the sheer raw power and energy of their sun would swamp any radio messages of theirs that might reach us. And if the aliens were trying to focus their communications as accurately as possible, they would certainly not be wasting their time focusing their signals at us.

My personal feeling, therefore, is that all attempts to detect radio signals from space originating from other intelligent life forms in the universe should be left to the private enterprise of those that do not believe me.

Thoughts on energy and the aether.

My main reason in discussing the sun in the previous section was not really to talk about the problems of radio communications between stars but rather to try to give one an idea of the vast amounts of energy being released by the sun each second and to perhaps get one to think about what is really happening to all that energy.

Lets think about all the light from the sun that reaches all the planets in our solar system. Mercury receives about the same amount of energy from the sun as the Earth does. Venus receives about twice as much energy from the sun as the Earth does. Mars receives perhaps a tenth of what the Earth does, Jupiter about four times that of Earth, Saturn about the same as the Earth, Uranus about two and a half percent of what Earth does and Neptune about one percent of the Earth. When one adds up all the energy that all the planets receive from the sun, one can see that in total all the planets and their moons receive about ten times as much energy as the Earth receives. In total this amount to about five billionths of the total energy that the sun emits.

All the rest of the energy produced by our sun is being radiated out into space. If one thinks about it, virtually all of the energy that has ever been released by the sun during the life of the sun is probably still travelling through the aether. And the same is true of all other suns in the universe. The aether itself is thus the place where virtually all the energy released not only by our star but by all other stars is stored.

Thoughts on the energy in nuclear reactions.

Once one starts to understand how it is possible for enormous amounts of energy to be present in what appears to be empty space, then one's whole perspective on the relationship between matter and energy changes.

In the last hundred years, since Madame Curie isolated radium and people started wondering why it glows and where the energy for it to glow comes from, humanity has been introduced to atomic power. We are slowly starting to understand the principles of nuclear fission and nuclear fusion. In both types of nuclear reactions, apparently very small amounts of matter have the ability to release huge amounts of energy.

The foremost question in my mind is where does all the energy released in a nuclear reaction come from. Current concepts suggest that matter is a form of energy and that matter is converted into energy in a nuclear reaction.

This does not, however, explain in a way that I can understand what energy is and what matter is and how matter can be converted into energy.

In Section 2 I proposed an alternative explanation, namely that energy is matter in motion. Using this concept I have been able to give an explanation for gravity. If my theory is correct, then what is the cause of the tremendous amount of energy released in a nuclear reaction?

My feeling, once again, is that one must look to the aether for the answer. For the aether is the place where energy is stored.

My feeling is that during a nuclear reaction, whether fission or fusion, conditions might be set up so that some of the matter taking part in the nuclear reaction becomes so concentrated for very short amounts of time that it interferes with the passage of the aether through the matter. This might cause part of the energy in the aether, that would normally be reflected from or pass through the body without hindrance, to be transformed into electromagnetic energy in a way that I cannot yet begin to explain.

In my derivation of the law of gravity, I assumed that some type of interaction between the aether and large bodies of atomic matter was taking place such that the large bodies of atomic matter reflected the aether and in doing so gained a small amount of momentum. The reason for this is that we know from everyday experience that while gravity can impart a certain small amount of momentum to us which can subsequently be converted into heat, we do not appear to be receiving any other massive amounts of energy from the aether other than that which can be explained by the force being detected as the force of gravity.

Imagines particles of the aether, travelling at speeds of at least 2.5E+10 kms/sec. The momentum that they transfer is proportional to their velocities. Their kinetic energy is proportional to the square of their velocities. When one is dealing with velocities of the order of 2.5E+10 kms/sec and then squares this number, the quantities involved are staggering. If some means could be set up to stop a little bit of the aether dead in it's tracks, then the gain in momentum that resulted would be the least of one's problems. The sheer amount of kinetic energy released would be enormous, far more than anything that we are typically used to dealing with or controlling in our everyday experience.

Is it conceivable that in all types of nuclear reactions, conditions are set up where a small part of the aether that would normally be reflected or pass undetected through a body is actually stopped dead in it's tracks, releasing huge amounts of kinetic energy which is converted into electromagnetic energy? This electromagnetic energy could be what we detect as the energy from a nuclear reaction.

If this is what in fact happens, then the energy produced in a nuclear reaction does not really originate from the matter that we regard as taking part in the reaction but rather from the aether itself. So just as an electric light bulb will glow as a result of electrical energy being continually fed to it, so could the sun glow as a result of energy from the aether being continually fed to it. And just as a light bulb is not the source of the energy being radiated from it, so might the sun not be the source of the energy being radiated from it. Rather the aether is the source of the energy being released by the sun. And if this is so and the sun is merely acting as a means of releasing energy already present in the aether, then the mass of the sun might be consumed at a rate far less that has hitherto been imagined. As in the case of a light bulb, where there is no relationship between the mass of the light bulb consumed and the energy released by the light bulb, so might there be virtually no relationship between the mass of the sun consumed in nuclear reactions versus the energy released by the sun. If this is so, then the sun might have existed for far longer than has been imagined until now. Is it perhaps possible that Jupiter, Saturn, Uranus and Neptune were at different times much closer to the sun, say in the position that the Earth is today, and were once similar sizes to the Earth? Could they have, each in turn, slowly drifted away from the sun over the course of tens of billions of year and gradually increased in size as they scooped up matter ejected from the sun?

One thing to bear in mind, however, is that matter in the sun is very definitely being used up very slowly. The reason for my thinking this is the way in which stars gradually fade out as they move further out from the center of a galaxy. It seems to me as if stars in the galaxy slowly consume galactic dust in their vacinity. As they move further out from the center of a galaxy, the amount of dust available gradually decreases and the stars gradually fade out as their supply of fuel drops.

I have been intrigued recently reading about two different types of experiments. One concerned gas bubbles that were made to collapse extremely rapidly. The other concerned fine tungsten wires that were heated with massive electric currents. In both cases, there is evidence that more heat was produced that can be accounted for. Is it possible that by merely compressing any type of atomic matter to a high enough density, one can set up conditions where some of the energy in the aether, which would normally pass undetected through a body, is in fact converted into energy that we can use?

Thoughts on how gravity could be affected by atomic energy.

If atomic energy is as a result of releasing a small amount of energy in the aether, then Newton's law of gravity needs yet another refinement. Consider the Earth moving around the Sun. Suppose that all the electromagnetic energy released by the Sun was as a result of converting a small amount of the energy in the aether into atomic energy.

If one considers the effect of gravity upon the Sun due to the Earth, this will not be affected by any aether used up by the Sun.

In the case of the Earth, however, the effect of the Sun upon the Earth is actually a little more than the effect of the Earth upon the Sun. The reason for this is that in addition to the fraction of the aether reflected by the sun, and thus prevented from reaching the Earth, aether is also being used up in the Sun to provide the electromagnetic radiation released by the sun. If gravity travels at 83,000 times the speed of light, then the mass of aether converted into electro magnetic energy is $1/83000^2$ or $1/6,900,000,000$ of the mass of electromagnetic energy released by the sun. The fraction of this total amount of aether used up that will not reach the Earth is, as was explained earlier, $4.5*10^{-10}$ of the total aether used up. This will further reduce the amount of aether reaching the Earth in the direction of the Sun to the Earth. This would have the effect of making the Sun appear to have slightly more mass than it actually has with respect to gravity but not with respect to centrifugal forces.

Thoughts on light and the aether.

For the last three hundred years, since Newton proposed his corpuscular theory of light, there has been a continual controversy as to whether light is corpuscular in nature or wave-like in nature. This matter has never been totally resolved. Maxwell, to my mind, established fairly conclusively that light is as electromagnetic phenomenon that travels in waves. In the early twentieth century the possibility that light is corpuscular in nature was once again raised by introducing the concept of photons to explain the opto-electric effect.

Quantum theory also assumes that energy consists of discrete units of energy.

The question to resolve, in my mind, is whether light is a wavelike phenomenon or a phenomenon based upon discrete units of matter, or energy, called photons or quanta.

As I mentioned in my propositions in Section 2, my feeling is that in the end all energy is simply matter in motion. This means that ultimately for energy to be transferred, matter must in some means be moved. So in the end, all energy is corpuscular in nature.

But this does not mean that light itself consists of massive particles moving at the speed of light throughout the universe.

As an analogy consider water. Think of the oceans and how energy is transmitted through the oceans. The oceans consist of vast numbers of water molecules all pressed down towards the Earth by gravity, i.e. the aether. If energy is suddenly generated in some part of the ocean, such as by an earthquake, then this energy is transmitted away from the source of the energy by the water molecules in the ocean. At the points where the oceans end, i.e. on sea shores, the energy will be released when it can no longer travel through water. At the points where the energy is released, there can be a significant and obvious movement of water. The thing to understand, however, is that although water can move rapidly at the final point where energy is released, the water that is moving at the points where the energy is finally released, i.e. along the shore lines, did not itself move from the original source of the energy, the earthquake. It is important to understand that the energy is not transmitted through the oceans by individual water molecules moving directly from the source of the energy until they reach the point at which the energy is released. Rather energy is transmitted through the water as shock waves by individual molecules of water impacting upon adjacent molecules of water and transferring the energy in this manner from one molecule to the next. If one considers the total average movement of water molecules in the oceans that transmit the force of an earthquake, it is probably very close to zero as each water molecule is initially pressed towards the next molecule and then pressed back towards it's original position after the shock wave passes.

The question to determine in the case of light, as in the case of gravity for that matter, is whether light is transmitted by individual particles moving directly from the source of light to the place where light is detected, as Newton suggested, or whether the transmission is rather one of some type of transmission whereby energy is transmitted from one particle to the next as it moves through the aether, in a similar way that shock waves are transmitted through water.

To my mind, light is very definitely not transmitted by means of particles moving from the source of light directly to the point where it is received. I feel that it is rather transmitted in an analogous fashion to the way in which energy is transmitted through water as a shock wave. The main reasons that I feel this are:

1. Light exhibits well documented wave like properties which cannot be explained if light moved as corpuscles.

2. Any phenomenon that works by means of particles moving directly from the source to the place where they are received would provide a number of clues to this fact as the distance between the source and destination increases.

    For if light was as a result of particles moving directly between the source and the destination, then one would have to assume that a particle, once it has been ejected by the source, would remain in the same state that it was once it left the source until such time as it struck the receiving object. In particular, one would not expect a single particle to grow in size or somehow split up into smaller particles moving in different directions after it left the source.

    This would then present a problem in terms of how light is radiated over very large distances. For irrespective of how rapidly or how many particles left the source, as they travel through billions of light years of space they would become more and more separated from each other. One would, therefore, expect that any view of items billions of light years away would exhibit a granularity due to the fact that particles from the source would have moved apart very widely over billions of light years of distance.

    But this does not happen. Light from the furthest distances recorded appears to behave in exactly the same way as light from nearby sources. This could only happen if light is transmitted by means of one particle of light in the aether transmitting light to adjacent particles.

3. If light were propagated as particles directly from the source to the destination, then one would expect a very different type of interaction when light from multiple sources pass through the same volume of space. For one would then expect particles of light to collide with each other and, at the least, to be deflected in their paths. This does not happen. The only effect that is known that disturbs the passage of light from the source to the destination is that large bodies, such as suns, cause the passage of light through the aether to be bent. No phenomenon, other than interference which is a wave like phenomenon, has ever been observed whereby light moving from one point to another interferes with or prevents light moving from a different point to a different destination. I feel very confident that if any such phenomenon did exist, we would be able to see it with our own eyes every day.

There is one very big difference present, however, in any analogy that we might draw between a shock wave passing through water and a light wave passing through the aether. In the case of a shock wave in water, the wave is passing at a greater speed through the water than the water itself ever moves at. In the case of a light wave, the wave of light is passing through the aether at a speed some 83,000 times slower than the speed at which gravity passes through the aether. I am still trying to understand how this happens. Once someone can explain this, I am sure that they will be able to explain what electricity and magnetism are. I will present some further analogies later as to what I think might be happening. But in the meantime, my personal feeling is that Maxwell was correct in his analysis of electromagnetic radiation.

Thoughts about uncertainty.

Any student of science these days will have heard about Heisenberg's uncertainty principle. My personal feeling is that in some respects there has been a lot of uncertainty about what is and is not possible to explain using this uncertainty principle.

A very clear distinction has to be drawn between the reasons for uncertainty as opposed to the way in which physical objects behave when uncertainty is involved. For in any experiments that we might perform, two different kinds of uncertainty could be involved, both acting at the same time but having very different causes and results.

The first type of uncertainty is that which I discussed in Section 1 and which I prefer to refer to as transmission delay rather than uncertainty. For there is nothing uncertain about transmission delay at all, once it is understood. As I discussed in Section 1, when any physical process is involved that takes a finite time to propagate, a type of uncertainty is involved. For by the time that a body is affected by the propagation of whatever is being propagated, the source of the propagation could have moved from it's original position and there is no way that the affected body can determine where the source of propagation actually is at the time that the affected body detects whatever is propagated.

As explained in Section 1, the Moon is gradually moving away from the Earth due to the transmission delay involved resulting from the finite speed at which gravity is propagated.

The fact that there is transmission delay involved in a process does not, however, mean that a body could ever behave in a way that defies the laws of physics. For, as explained in the beginning of Section 1, all bodies will behave exactly as predicted by physics on the basis of where all bodies involved appear to be. This will cause behavior that might, until the cause of the transmission delay is determined, seem to defy the laws of physics. But once the reason for the transmission delay is understood it should be possible to explain the behavior of the body exactly according to the laws of physics.

There is a second type of uncertainty involved in any experiment that we might perform, namely that we might not be ever able to exactly measure the motion or position of a body taking part in the experiment. This kind of uncertainty is quite different to transmission delay. For transmission delay results in bodies acting in different ways than they would if no transmission delay was involved, i.e. if propagation of forces took place at infinite speed. This second type of uncertainty, due to our inability to measure something accurately, has no effect whatsoever upon the way in which the bodies in experiments actually behave as the bodies in the experiments could not care less whether or not we are trying to measure their behavior. This does, of course, assume that the way in which we are trying to measure something does not in itself interfere with what is being measured. This second type of uncertainty can in itself make it very difficult to understand experimental results at times.

The real difficulty arises when we have both transmission delay as well as uncertainty in measurement involved at the same time. In such cases it can become very difficult to understand the results of an experiment. In order try to fully understand the results from such experiments, one must be very careful to try to separate the results of transmission delay from those due to uncertainty of measurement

To the extent that Heisenberg's uncertainty principle is valid in developing mathematical equations to predict certain experimental results, my feeling is that it is in fact dealing with uncertainty that is as a result of transmission delay. Two types of transmission delay might be involved. The one is that moving at the speed of gravity, concerned with gravitational effects. The other, more common one, is due to transmission delays involved with electromagnetic effects moving at the speed of light.

I feel very strongly that there is nothing in the universe that would ever behave in an uncertain manner. All bodies behave exactly according to Newton's laws of motion once account is taken of transmission delays. My personal feeling is that whenever confusing experimental results are obtained that seem to defy the laws of physics, our first reaction should be to accept that the real reason for the uncertainly is most likely to be our uncertain understanding of what is actually happening rather than due to any uncertainty in the physical processes actually involved. And I feel that in such cases, the first approach should be to try to understand what is actually happening rather than to simply try to develop an empirical equation that models what is happening.

Thoughts on quantum theory.

If the energy of light is transmitted in an analogous fashion to shock waves through water, then certain concepts in quantum theory need to be re-examined.

Before considering the question of light itself, I would like to present two analogies to try and indicate some of the concepts that I feel are important in understanding how light might work.

Have you ever been in bed at night and heard water dripping from a faucet (or tap) into a basin? Lets think about what it really happening when we hear that simple drip, drip, drip sound. Firstly, water is slowly dripping out of the faucet as a series of drops of water. The water then falls through the air and hits the surface of the basin. This impact of water on the surface of the basin causes a drip sound to be generated. The sound is then transmitted through the air towards our ears. Once the sound reaches our ears, our ears then reflect the sound around inside our ears through most complicated ear canals towards our ear drums. Once the sound reaches our ear drums it causes our ear drums to vibrate and generate electrical impulses which are then transmitted by auditory nerves to our brain which finally, in ways that we do not understand at all yet, enables us to hear the sound.

Let us think about what this simple analogy can teach us. First of all, the cause of the dripping sound is that water under pressure is slowly passing through the valve in the faucet. After passing through the valve the water flows downwards towards the bottom edge of the faucet. At that point it joins other water molecules already at the edge of the faucet and forms a drop which gradually increases in size. The water forms a drop due to surface tension in the water. When the drop reaches a certain exact size the weight of the water in the drop finally reaches the point where the surface tension in the water can no longer sustain the weight of the drop. At that point the drop of water falls from the edge of the faucet and moves down through the air towards the basin. If we just think about the way in which water falls to the basin some interesting facts emerge. Firstly, although a drop of water consists of billions of billions of water molecules, the water flows from the faucet in a series of drops of fairly uniform size. It is almost as if the energy of the water is being released in discrete units, or quanta, being the potential energy of a drop of water falling through a distance from the faucet to the basin. The important thing to understand is that although the quantum of energy in each water drop is fairly constant, the water drop itself is sub divisible. The second thing to understand is that water does not always flow in quanta of a drop at a time. For the water flowing through the valve in the faucet, due to the high pressure of the water at the inlet side of the faucet, is being squeezed through the valve almost continuously in much smaller quanta, or units, of water. This can easily be demonstrated by observing that the individual drops at the edge of the faucet gradually grow in size until they eventually fall from the faucet. What is effectively happening is that potential energy is gradually accumulating in the water drop as the size of the water drop at the edge of the faucet gradually grows. The water drop is held at the bottom edge of the faucet in a delicate balance between gravity and surface tension. When the force of gravity exceeds the maximum force that surface tension can sustain, the water drop falls from the edge of the tap. This example illustrates, therefore, a situation in which energy is being released at regular intervals in discrete quanta due to the fact that a situation exists whereby something prevents energy from being released until a certain threshold has been reached. In this case, the limiting factor is surface tension.

Let us now think about what happens when a water drop falls from the faucet and hits the surface of the basin. As the water drop falls through the air, gravity accelerates the water drop converting energy in the aether into kinetic energy in the water drop. When the water drop reaches the surface of the basin it is stopped in it's tracks by the basin. At that point the kinetic energy gained by the water drop is released. Part of this energy is released as the dripping sound. Due to the fact that all the water drops are fairly uniform in size, the amount or energy of sound generated by each water drop is also fairly similar to that of any other drop. So at this point we have sound being generated in quanta of energy which are quite different from the quanta of energy that were present in the water drops themselves at the time they left the faucet. Once the sound is generated it is then transmitted by the air towards our ears. Now we know that the air itself consists of billions of billions of molecules of oxygen, nitrogen, water, carbon dioxide, etc. moving randomly through space and colliding with each other. This air is under pressure, once again, due to gravity. We know that the velocity of sound in the air depends upon the air pressure and type of gas. The important thing to understand at this point is that even though the sound energy from the water is released in quanta depending on the size of the water drops, the energy once released is transmitted through the air in all directions. The original quantum of sound energy is evenly distributed by the air as it moves away from the point of impact. There may well be a minimum quantum size of sound energy that could be transmitted through the air. One would, however, expect this quantum size to be billions of billions of times less than the amount of energy released when a drop of water falls. Another important thing to understand is that although the water falls in quanta of drops, the dripping sound of the water drops is transmitted in a totally different medium by a gas where the molecules in the gas are very much smaller than the quantum size of a drop of water and where the molecules in the gas transmit the sound at a much higher speed than the drop of water itself falls. And the final thing to understand in this analogy is that three different "aethers" coexist in the same space. For sound is transmitted through a medium, the air, in fairly slow moving and low frequency waves. It is also possible, however, to see the water drops fall as light is transmitted through the same volume of space by a totally separate medium, moving at a much higher speed and with a much higher frequency. And, if one thinks about it, gravity is also passing through the same volume of space, moving at an even higher speed in an as yet unknown manner, either as waves or particles.

A second analogy might also help the reader to understand what I am trying to explain.

In November 1940 a bridge in The United States of America, called the Tacoma Narrows bridge, collapsed. I would encourage anyone interested in understanding light to examine the video of that bridge collapsing. What happened was that on that particular day a wind of over forty miles an hour started blowing across the bridge. It is my understanding that the wind was blowing at just the right speed to cause resonance in the bridge. The bridge, over a period of a couple of hours, slowly accumulated energy and began swaying and oscillating more and more violently until sufficient energy had been accumulated at which point the bridge broke and fell into the river.

There are a number of things to understand from this analogy that are, I feel, quite important in understanding the behavior of light. The first is that the bridge had been standing for many months before it broke. During that time it had been given the name "Galloping Gertie" due to the way it used to behave in the wind. Prior to the collapse, however, nobody expected the bridge to collapse. What was needed to break the bridge was a wind to be blowing at just the correct speed to cause resonance in the bridge. The second is that the bridge could have easily withstood wind gusts at a much higher speed without breaking. The third thing is that the amount of energy transmitted from the wind to the bridge in a short period of a couple of minutes would not have been sufficient to break the bridge. The thing that broke the bridge was that when the wind blew at just the right speed it set up the conditions whereby small amounts of energy from the wind could gradually be transferred to the bridge and, more importantly, be accumulated by the bridge. This caused the bridge to start oscillating in a way that enabled the small amounts of energy from the wind to be gradually stored up inside the bridge as the amount of oscillation gradually increased. It was only after a couple of hours, when sufficient energy had been accumulated, that the bridge eventually broke and released all the stored up energy in a mighty crash.

At the time of Maxwell's death in the 1800's, it was widely accepted that light was transmitted as waves through the aether. Early in the twentieth century this idea was challenged to explain the photo-electric effect. The concept of photons was introduced to try to explain this effect.

This has resulted in a sort of stalemate being reached whereby light is assumed to move as waves under certain circumstances and as particles or photons under others.

As mentioned above, my feeling is that light moves as waves through the aether. How can one reconcile this view with that whereby light is assumed to be photons that have energy that is proportional to their frequency.

To my mind the first point of confusion has been a lack of understanding or appreciation of a number of different steps involved in the propagation of light. For there are really three separate and distinct steps involved in the transmission of light.

The first step involved is when light is initially generated and released by an atom or molecule.

The second step is where light is transmitted through the aether.

The third step involved is when light strikes an object.

The three steps are quite separate and independent of each other.

Let us consider the first step, where light is initially generated and released by an atom or molecule. This is, in a way, analogous to a drop of water breaking free from the edge of the faucet or the bridge breaking or an earthquake caused by land suddenly slipping. In all the analogies mentioned, items such as a drop of water, a bridge or a land fault were slowly accumulating energy until such time as some sustainable limit is exceeded. At the time the limit is exceeded, all the accumulated energy is suddenly released in a very short time as a burst of energy. My feeling is that atoms and molecules can, in an analogous way, slowly accumulate energy until certain energy levels are reached at which point energy is suddenly released in bursts, or quanta.

This would, to my mind, explain a number of different things known about atoms and light. Firstly, light would very definitely be emitted in discrete units of energy, knows as photons. In addition, different elements would be expected to have different energy levels at which energy is released. This is what happens in practice.

Let us consider the second step where the light energy released by an atom or molecule is transmitted through the aether. This is analogous to the dripping sound of water being transmitted through the air or a shock wave from an earthquake being transmitted through the ocean. Once the dripping sound of water has been generated or an earthquake has occurred, the energy is released into the transmitting medium, i.e. air or water, and what occurs after that in the medium is totally independent of the initial cause of the disturbance in the medium or, for that matter, the initial amount of energy released. In a similar way, when light energy is released into the aether to be transmitted through the aether, what happens to light in the aether is quite independent of the atom or molecule that originally released the energy or the amount of energy that was released.

In the same way that once the quantum of energy in a drop of water is released, or the quantum of energy in an earthquake is released, so once a quantum of light energy from an atom or molecule is released, it is then dispersed throughout the medium in an almost continuos fashion where the energy can be subdivided almost continuously by the medium over vast distances. In the case of a drop of water falling, the energy once released into the air as a dripping sound is dispersed by the air and broken up into very much smaller quanta that would depend on the size of air atoms, as opposed to the original quantum of energy which was released which depended upon the maximum size possible for a drop of water. In the case of the energy released by an earthquake, it is gradually dispersed by the ocean until it reaches thousands of miles of coastline around the world. In the case of light, the energy once released by an atom or molecule is dispersed by the aether as it travels further and further from the source until it strikes an object. It is important to understand that although the atom or molecule releasing energy might be releasing it in discrete units, or quanta, of energy, once this energy has been transferred to the aether the aether will then break this quantum of energy down into very much smaller almost continuous units. There is experimental evidence that very small amounts of light from a source, certainly less than a quantum, can be made to interfere with each other. This proves, to my mind, conclusively that the quanta of energy which are released by atoms can be broken up by the aether and can cause predictable wave like behavior in units of much less than a quantum.

Let us now consider the third step in the transmission of light, namely when light strikes an object. As is well known, one of three things can happen. The light can be reflected from the object, the light can be absorbed by the object or the light can pass through, or be transmitted through, the object. Let us consider the most important case where light is absorbed by an object. A tremendous amount is known about this. To my mind, however, the most important thing to consider is that even though atoms and molecules release energy in terms of quanta, they must be able to absorb energy in quantities very much less than a quantum. For, as has been explained, a quantum of light once emitted by an atom or molecule is gradually dispersed in the aether as it travels further and further from the source. Consider an atom emitting one quantum of light. If this light is emitted in all directions then any other atom or molecule that received any of this quantum of energy could never receive the entire quantum of energy originally emitted. For at least half of the energy would have been radiated away from the atom in the direction opposite to the atom that received energy. If one thinks about the implications of this, what this really means is that if an atom or molecule is capable of absorbing light energy then it must be able of absorbing this energy in quantities very much less than the quantum amount of light which it itself might at a later stage emit. For most of the light energy reaching any single atom or molecule would be arriving continuously from all directions in quantities very much less than a quantum of light. In addition, whereas a particular frequency of light might be generated by only one particular element or molecule, once this energy is released at the particular frequency, it can be absorbed by a large number of different types of elements and molecules, none of which may generate light at that particular frequency. This once again indicates that the rules for absorption of light are quite different from the rules of generation of light. If we think of the drop of water slowly growing in size until it reaches a certain size and falls, or the bridge gradually accumulating energy until it breaks and releases all the energy at once, in both cases we have a situation where energy is being slowly accumulated until a point is reached at which it is all released at a single time. In both of these cases the effects observed could not happen unless the drop of water and the bridge were capable of absorbing energy in much smaller units than it was released in. In the same way, it would not be possible for molecules to slowly accumulate energy and then release it in discrete units unless they were capable of absorbing much smaller quantities of energy than they release.

And, like the bridge, atoms might be more inclined to absorb energy which was at just the correct frequency to cause oscillation within the atoms. It is well know, from atomic absorption spectroscopy, that different elements absorb light at very specific frequencies dependent upon the types of elements involved.

Thoughts on the photo-electric effect.

As discussed above, my feeling is that any theory that attempts to propose that light is transmitted through the aether in terms of the same quanta, or photons, as light is originally emitted by atoms is bound to fail. How then, does one explain the photo-electric effect?

The photo-electric effect is very simply an effect noticed where certain substances have the property that when light falls on them they have the ability to generate an electric voltage and electric current. One of the peculiarities noted is that the voltage is generated in these substances depends upon the frequency of the light falling on the substances. If light is below a certain frequency then no voltage is generated. Once light is above a certain frequency, then a voltage is generated and the total quantity of electricity generated is then dependent upon the total quantity of light falling on the substance.

The important thing to understand in the photo-electric effect is that we are dealing with what happens in the third step in the transmission of light. Namely the final stage where light falls on a body and is absorbed by the body.

The concept proposed to try to explain the photo-electric effect by means of photons by it's very nature assumes that a body is absorbing energy in units of photons. This means that once a photon of light is emitted by a substance it then has to pass unchanged through the aether until it strikes the receiving object. I have stated my objections to this concept.

The real problem is how does one explain that only light above a certain frequency can generate the photo-electric effect. I would ask the reader to think once again about the example cited earlier of the collapse of the Tacoma Narrows bridge. As mentioned earlier, the bridge fell because wind blew across it at just the correct velocity to cause resonance in the wires of the bridge and allow the bridge to slowly absorb energy from the wind.

There is a great deal of evidence to suggest that different elements and molecules have the property of absorbing light radiation at very specific wave lengths. Each thing that we look at has it's own color which is as a result of the atoms and molecules on the surface of the substance absorbing certain wave lengths of light in preference to other wave lengths which are reflected. It is well known that any element will absorb very specific wave lengths of light. This property is in fact used in atomic absorption spectroscopy to analyze the presence of specific elements in samples.

The whole question of how light is absorbed, reflected or transmitted through objects is obviously a very complex one. Two fundamental ideas in all considerations of what happens when light strikes a body are, to my mind, that all bodies have the ability to absorb light energy in units very much less than the quanta which they might radiate light at and that each body, for reasons unknown, has the peculiarity that it will absorb certain frequencies of radiation in preference to other frequencies.

The fact that the photo-electric effect only occurs in certain substances above a certain frequency of light indicates to me nothing other than that those photo-electric substances preferentially absorb higher frequencies of light and do not absorb lower frequencies. To my mind the problem is really nothing more than one of trying to explain why certain substances are red and others are green or yellow or blue. There is something going on within all atoms and molecules that gives them the ability to preferentially absorb light of specific frequencies. In the case of photo-electric compounds, one obviously would not expect any effect to be present for light with frequencies that are not even absorbed by the photo-electric compounds. For the whole process must surely be at least a two step process where first energy is absorbed by the compounds and then secondly used to generate the photo-electric effect.

Those familiar with radio waves, particularly those who have built their own radio receivers, fully understand how it is possible, by simply using a capacitor and an inductor in a circuit, to build an oscillator and how, by changing the capacitance or the inductance, one can tune the receiver to be particularly receptive to a very narrow range of frequencies. This is, to my mind, a very simple way of understanding the reason why a circuit, or body, could be inclined to absorb electromagnetic energy at very specific frequencies. There is, to my mind, no need to get into any concepts of photons to understand the photo-electric effect.

Thoughts on Michaelson and Morley's Experiment.

Michaelson and Morley, over a hundred years ago, performed their famous experiment to determine how fast the Earth is moving through the aether. This experiment, to the surprise of Michaelson and Morley as well as many others, found no evidence that we were moving through the aether at all.

This famous experiment lead to concepts of the speed of light being constant with respect to any frame of reference. This in turn lead to the concept that the rate at which time progresses varies as a function of one's speed. It has also lead to the concept that nothing can travel faster than the speed of light. It is my understanding that it has now been found experimentally that atomic clocks aboard satellites run at the same speed as those on the Earth and do not experience any time variation with respect to their speeds. I have also suggested that gravity is propagated at about 83,000 times the speed of light. I will leave it to others more capable than I am to work out the contradictions in these theories.

The fallacy in Michaelson and Morley's experiment is that Michaelson and Morley started off by making the same incorrect assumption that Newton did, namely that the aether is somehow stationary and that it is the Earth which has a greater velocity than the aether. Michaelson and Morley were thinking of the Earth moving through the aether in a similar way to the way in which a bullet flies through the air.

My personal feeling is that the real explanation of why Michaelson and Morley's experiment gave the results that it did is very simply that it not the Earth that is moving quickly through a stationary aether. Rather it is that the aether itself that is moving very much faster, in all directions, than the Earth and in fact moving through the Earth and is in fact actually moving the Earth.

Consider the fact that gravity is transmitted at a speed of over eighty thousand times that of light itself. As I mentioned in my theory of gravity, we are pushed towards the Earth due to very small imbalance in the aether caused by the Earth preventing a very small amount of the aether that would have reached us in the direction from the Earth to us. In a similar way, the Earth is pushed towards the sun by an imbalance in the aether. For the universe to be in the equilibrium that it apparently is, we must assume that the aether is very nearly in equilibrium in all directions. If it were not, we could expect that, over the course of billions of years, our galaxy as well as the sun and the Earth would have been accelerated by the aether to the point at which the aether was in equilibrium with them. In either situation, the aether would appear to us to be in equilibrium.

What we are really dealing with, therefore, is an aether which is moving through us in all directions in almost perfect equilibrium and where gravity is being propagated at speeds of 25 billion kilometers per second. The speed of the Earth around the sun is at most 29 kilometers per second. In relation to the speed of gravity this is nearly a billion times less. The speed of our sun moving around our milky way galaxy is probably a few hundred kilometers per second at the most.

I suspect that light is transmitted by some type of "atmosphere" of particles which are larger and less dense than those of the particles of pure matter and which move at a slower velocity to that of the particles transmitting gravity but still move at a speed at least that of light. These particles themselves form a kind of atmosphere in a similar manner to that of our atmosphere. These light particles must move at least as fast as the speed of light in order to enable light to be propagated at the speed of light. These light particles themselves are under pressure due to the aether itself.

If such a situation exists, then given that we must be in equilibrium with the basic aether, then so must the larger particles which comprise the medium which distributes light also be in equilibrium with the aether. One can get a very good idea of the fact that the medium that transmits light is very close to equilibrium by observing that light is bent by gravity as it passes by large objects. We can detect hardly any effect of light being changed by the gravity of the Earth apart from some experiments that I read of where the wave length of light changes very slightly as light moves towards or away from the Earth.

If that is the case, then the most that we would be out of equilibrium with the aether would be perhaps a few hundred kilometers per second in an aether transmitting gravity at a speed of 25 billion kilometers per second. But even then, I would argue that we would be in equilibrium with the material in the aether responsible for transmitting light as the difference in pressures of the aether is forcing our sun and the Earth in an orbit of some type about the core of the milky way galaxy. The reason an astronaut feels weightless in orbit about the Earth is because he is offering no resistance to gravity caused by an imbalance in the aether. I would argue that in the case of the sun, the Earth and whatever it is which comprises the medium transmitting light, that all of these things are offering no resistance to any imbalance in the aether caused by the core of the milky way galaxy and any other large bodies which might be affecting them. In a similar way in which an astronaut feels weightless, I would expect there to be no measurable effect between us and whatever it is in the aether acting as the medium for light caused by us moving around the galaxy. But even if I am wrong about this, my personal feeling is that one could hardly expect that Michaelson and Morley's experiment would detect any change in wave lengths other than perhaps a few hundred parts in twenty five billion.

Thoughts as to whether the Universe is in Equilibrium.

One of the most important questions to answer, to my mind, is whether the universe can remain in a permanent state of equilibrium over billions or trillions of years.

Current thermodynamic theories suggest that, in any transfer of energy, entropy must always increase. Based on this theory alone, the universe can never be in a state of perpetual equilibrium as all types of energy transfers would result in an increase in entropy. For the universe to remain in a permanent state of equilibrium that might have existed for trillions of years, some means would have to exist whereby the total entropy in the universe would not increase.

I would like to ask all scientists to once again put aside all their previous knowledge learned according to current theories and examine openly the possibility that it might be possible for the total entropy in the universe to remain constant.

Based upon current theories, the Moon should not be able to move away from the Earth. And yet it does. I have explained in Section 1 that this is due to the finite speed of gravity. Based upon our current theories, when a body drops towards the Earth due to gravity there is an increase in entropy. On this basis, when the Moon moves away from the Earth, this really should be a decrease of entropy. As I explained in Section 1, the movement of the Moon away from the Earth is due to the transmission delay introduced due to the finite speed of gravity.

What the Moon moving away from the Earth demonstrates is a practical example whereby entropy can apparently be decreased due to transmission delay. The thing that I would like scientists more capable than me to examine is whether this is not a simple example of a more general rule, namely that whenever transmission delay is involved, entropy as defined by our current theories and equations can decrease.

I discussed, earlier in this section, the possibility that light can slowly loose energy as it passes through the aether. I also mentioned that I would like to discuss where this energy might be going to. In order to try to explain what I am thinking, I will try to use a couple of analogies to illustrate some of what I feel are the more important items to be considered.

Consider an example of a container filled with a gas, such as oxygen. Now we know from thermodynamics that for a certain quantity of gas at a given volume, the pressure of the gas depends upon the temperature. Suppose that we could have a vessel containing a fixed amount of oxygen at a certain temperature that is perfectly insulated in that heat can neither enter or leave the vessel. We would be justified, under current theories, in then being able to assume that the temperature of the gas in the vessel would remain constant. Now, if the temperature, volume and mass of a gas are held constant indefinitely, then we expect that the pressure will remain a constant indefinitely. But we know that the pressure of a gas in a vessel is due to the random movement of the molecules of the gas. Hence, if the pressure were to remain constant indefinitely, one is inevitably led to the conclusion that in a perfectly heat insulated vessel the total net movement of molecules trapped within the vessel would remain a constant indefinitely. What we really have, therefore, is a situation where individual molecules are continually colliding. Any particular molecule could over the course of time and innumerable collisions with other molecules, be continually changing it's direction and velocity. But, over the course of time, the total net momentum of all molecules in the container in all directions would be zero with respect to the containing vessel.

Certainly, in any experiment which we might perform over a short period of time, we would expect, as we strive to minimize the amount of heat transfer to and from the vessel, that the pressure would remain constant.

But let's think about what could happen to this vessel if we were to continue the experiment for billions of years. Even if we could have a perfectly heat insulated vessel, other factors which we typically tend to ignore in our everyday experience would, over a period of billions of years, make their presence known and could no longer be ignored. For example, the some of the oxygen atoms in the gas would be different isotopes of oxygen, all with their own half lives. Some of these atoms would, it the course of billions of years, decay into other elements. This would inevitably generate a certain amount of heat which could not escape. This radioactive decay would also result in certain oxygen molecules that previously existed no longer existing. In addition, other elements would be formed. All of this would result in the total number of oxygen molecules in the container changing and the pressure would therefore change. What I am trying to illustrate is that if we were to examine a perfectly insulated vessel over billions of years there could be various factors involved which could cause a gradual change of pressure in the vessel over the course of that time. And none of these factors would be easily measured during the course of an experiment over the span of our lifetimes.

Think of another example of a shock wave generated in the ocean by an earthquake. Now we know, from experience, that the energy in this shock wave is transmitted through the ocean until it reaches land, at which time a tidal wave would be generated. Let's imagine that we could increase the size of the oceans tremendously so that instead of a shock wave travelling through the water for hours, it would take hundreds of years to cross an ocean. Under those circumstances, factors which we normally ignore when considering a tidal wave could no longer be ignored. For example, energy loss through friction within the water would become more of an issue. In addition, during the passage of the shock wave through innumerable water molecules, some of the water molecules would be on the surface of the water. Some of these molecules might actually evaporate during the brief period when they are actually transmitting the shock wave. Under such circumstances, a small amount of the energy of the shock wave would actually be lost due to evaporation. As the total time that the shock wave traveled through the ocean increased, so would the loss due to evaporation. Once again, we have a situation where a process that we can fairly accurately predict, based upon our experiences with tidal waves, would become very different if the time span over which the event took place increased enormously.

Now lets think about another type of energy propagation, namely electromagnetic radiation, or light. In all of our experiments with light where we do something with light, such as generating it, reflecting it, refracting it, making it interfere with itself, polarizing it etc., the light in our experiments typically travels for less than a millionth of a second during our experiments and, in the case of experiments involving space craft, a maximum of a few hours. It is highly unlikely, therefore, that we could ever expect to measure properties of light which might be present but which cumulatively only make their presence known as light travels through the aether during the course of millions or billions of years. So just as small effects, such as radioactive decay of a gas in a perfectly insulated vessel or evaporation of water in an infinitely large ocean, which we cannot measure in our experiments could become significant over billions of years, so could almost negligible but cumulative effects in the transmission of light become significant issues during the course of millions of years as light travels through the aether.

My feeling is that at the moment we cannot discount such possibilities simply because we cannot measure them in experiments lasting a few millionths of a second. If, however, by observing light that has traveled for millions of years, we have experimental evidence that some process appears to be taking place on a consistent and regular and predictable manner, then I feel that we would be foolish indeed to reject the possibility that such a process is indeed taking place on the mere basis that we have never been able to measure such an effect in the course of experiments lasting a few millionths of a second in our laboratories.

As I discussed earlier, there is irrefutable evidence that the wave length of light reaching the Earth slowly increases as the distance over which it travels increases. This relationship is known as Hubble's Law. I suggested earlier that this might be due to some type of transmission delay involved in whatever mechanism is responsible for the propagation of light.

As mentioned earlier, my feeling is that light could well gradually loose energy as it passes through the aether and that this is responsible not only for the red shift but also for the background radiation which is present in all directions. The big thing to understand, however, is that the background radiation is merely light which has lost such a large proportion of it's original energy that it's wave length has increased to the point where it is no longer recognizable as light from individual stars.

The background energy, by itself, cannot however account for where all the energy that was originally radiated from a star has gone to. The simple reason for this is that the total background radiation that we can measure represents the sum of all the electromagnetic energy that remains from light that was originally emitted by countless stars beyond the visible range of our universe. So far from representing all the energy of light that has been lost, the background radiation represents what is left of light from distant stars where most of the energy has been lost.

The question to ask is where has this energy lost by light as it travels through the aether gone to.

My belief is that this lost energy must have actually been slowly transferred into regenerating the energy stored by the aether itself.

As an analogy, think of the perfectly insulated container of gas. The molecules in this gas are continually moving and colliding with each other. At some points some of the molecules, in colliding with each other might increase their speeds while others would decrease their speeds. On average, however, over a long period of time an equilibrium would be established whereby the total net momentum of all molecules in the gas would remain a constant.

Is not the same sort of thing possible in the universe? If my theories are correct, then at the lowest level everything consists of little pieces of pure matter moving about randomly, colliding with other pieces of pure matter. In any single collision between two particles of pure matter the total momentum remains a constant and the total energy must remain a constant. Hence when one sums up all of the collisions taking place continually within the universe, the total momentum and total kinetic energy of all matter in the universe must remain a constant.

As has been mentioned, gravity is caused either by small particles in the aether or by some type of wave motion which in turn is caused by very much smaller particles in the aether pushing much larger bodies towards each other. This results in the high speed small particles giving up a very small amount of their momentum when they are reflected by much larger, much slower moving, aggregates of small particles. If, as I also suggested, nuclear energy is some type of process taking place whereby the energy of high speed small particles in the aether is converted into electromagnetic energy, then this too could result in some of the high speed small particles giving up most of their energy to larger, slower moving, aggregates of small particles which move in some way to produce electromagnetic phenomena.

Just as it is possible for two objects to collide in various ways that will either increase of decrease the speed of the colliding bodies, so should it be possible for particles of pure matter to collide occasionally in just such a way that will increase the total speed of part of one of the bodies. For example, most collisions will occur when two bodies strike each other. However, once in a while, three bodies all moving in different directions might occasionally collide with each other all at the same moment in time. This might result in most of the three bodies combining with each other and moving at a slower speed while a small part of the body moves away at a much larger speed. If one studies, in slow motion, what happens when a drop of water hits a surface of water, one will see that although most of the drop of water is stopped by the water, a very beautiful spray can be formed where a very small amount of water is shot upwards in the reverse direction to which the drop was travelling.

If this is the case then we could have a situation where there could be a permanent state of equilibrium over a tremendously long, if not infinitely long, period of time in the universe.

What I am suggesting is that there could be a very long term cycle in the universe where small particles of matter, moving at least the speed of gravity and perhaps very much faster, collide with very much larger particles moving at much slower speeds. These collisions result, at some level, it the effects known as gravity. As I also mentioned earlier, I feel that at some higher level, nuclear energy might be the result of interactions between the aether and small but quite dense particles which might be found inside a sun.

In both cases, namely gravity and nuclear energy, "high speed" energy from the aether, which is used to produce the properties of gravity and the generation of nuclear energy, would be "used up" and there would be a slow but steady drop in this type of total kinetic energy stored by the aether. In the case of gravity, the kinetic energy is used up modifying the speed or direction of bodies. In the case of nuclear energy, kinetic energy in the aether is converted into electromagnetic energy or heat when energy from the aether is released in what we term nuclear reactions. This electromagnetic energy would then be radiated into space.

My feeling is that over the course of billions of years of travel through space, the process by which light is transmitted is not perfect. A very small amount of the electromagnetic energy in the aether is gradually lost due to some type of transmission delay, such as the random collisions with other particles that result in smaller particles being formed at much higher speeds, thus replenishing the aether for the "high speed" energy which was originally lost by the aether when electromagnetic energy was created.

I therefore feel that there is a very long term, stable situation of equilibrium present in the universe whereby most, if not all, of the energy in the universe is stored in the aether. At some point, whenever this aether interacts with large bodies of matter, a very small part of this energy in the aether is converted into kinetic or electromagnetic energy. All this energy is ultimately radiated from the body as electromagnetic radiation. This electromagnetic radiation then travels through the aether and in the course of travelling through the aether over billions of years, with billions of billions of collisions taking place, ultimately transfers it's energy back to the aether again, thereby replacing the energy which was originally lost by the aether.

Final thoughts on the aether.

I would like to try to recap some of my thoughts on the aether which I have not, in my opinion, explained very well.

As I mentioned in my propositions in Section 2, I feel that all energy is matter in motion. I also suggested that matter itself has no properties other than that of inertia, as explained by Newton, and that two separate pieces of matter cannot be in the same place at the same time.

I have further suggested that at the lowest level the universe consists of nothing more than vast amounts of extremely small particles of extremely dense matter moving through the universe randomly and colliding with each other occasionally.

I have further suggested that all collisions between matter obey the laws set out by Newton. I further suggested that out of this chaos of random collisions, all the order that we witness in the universe grows as a result of the laws of probability.

I have further discussed that, as a result of the laws of probability, not all of the particles will move with similar velocities. What I have suggested is that there are laws of probability which result in larger bodies tending to move with lower velocities and smaller bodies moving with much higher velocities.

I feel that as we move from the very largest bodies that we know of, namely galaxies, to the smallest bodies, namely particles in the aether responsible for gravity, the speed of the bodies increase as the bodies get smaller.

I suggested that whenever we find bodies which tend to be of a fairly narrow range in sizes, such as galaxies, stars, planets, hail stones or snow flakes, atoms, protons and neutrons, quarks, etc, that we can be reasonable sure that these bodies are in turn composed of a large number of much smaller bodies and that there are limiting conditions which statistically prevent any type of body from exceeding a certain size.

Newton's work enabled us to understand to a large extent the motion of larger slower moving bodies. To some extent scientists have been working in a "top down" method of discovery in terms of applying Newton's laws to smaller and smaller bodies.

At present we do not have any idea as to what the cause of electrical charges, magnetism or electromagnetic radiation is. I have suggested a mechanism which could explain what gravity is. If one thinks of it, to some extent my postulates suggest a "bottom up" method of investigation.

My feeling is that at some point if we can continue down with our current "top down" method of discovery and also continue upwards with my suggested "bottom up" method, we will find an explanation as to what the mechanism causing electricity, magnetism and electromagnetic radiation actually is.

Newton, in the Scholium in his Principia, suggested that we might never find a cause for gravity, magnetism and light and suggested that what was known in his time was sufficient. I disagree with Newton on this point. Firstly, I do not believe that it is ever enough to merely know how to predict how something behaves without understanding why it behaves in the predicted way. If one has an equation that can apparently predict something very accurately but does not know why the equation works, then one is always at risk of incorrectly using the equation in situations where the equation does not apply. To give an example, once one has an idea of what might be causing gravity, it then becomes possible not only to derive Newton's equation of gravity but to more importantly suggest conditions under which the equation will and will not work. Secondly, once one has an idea of what could be causing an effect, one can then start predicting possible new ways to manipulate the effect to one's advantage.

In Newton's time, the knowledge and methods that Newton supplied to the world were so enormous in comparison to what had existed before him that it was sufficient for some centuries to guide mankind in our discoveries. My feeling, however, is that no matter how revolutionary a new idea may be, it will only be sufficient for a certain period of time until technology has caught up with the idea and it once again becomes important to develop a better understanding of what is really happening.

I gave the example, earlier on, of one lying in bed listening to water dripping from a faucet. Imagine that some inquisitive soul listening to the sound of the faucet dripping had the means to record the dripping sound on a computer and analyze it while lying in bed. It is quite conceivable that he could, using Fourier series or a number of other mathematical methods, develop a mathematical model to predict the sound. But if he did not know the reason why the sound was being generated, he would, in my opinion, be no closer to understanding what was really happening than a child who might be listening to the sound in the next room. In fact, if the child got out of bed and investigated the sound and saw the water dripping, the child would, without the use of any mathematical modeling techniques, have a far better understanding of what was actually happening than the person with his computer and mathematical models. In fact, if the child merely opened the bathroom door a little further to investigate the sound, he could change the air flow in the bathroom and thereby cause the drops to fall at a slightly different rate. This would result in the person lying in bed with his computer having to modify his mathematical model without knowing why. And if the child, finding the cause of the water dripping, turned off the faucet to stop the dripping altogether, the person working with mathematical models alone would be at a total loss.

My personal feeling, therefore, is that a mathematical model is no substitute for true understanding. In order to understand physical principles, one must get out of bed and investigate the problem. I have always felt that any person whose goal in life is to understand what is happening by means of developing the perfect equation is somehow missing the mark. I would urge anyone who wishes to understand what is actually the cause of light, electricity and magnetism to try to picture in his mind what could be happening, using analogies with other things that one understands, rather than starting out with the goal of developing a mathematical equation. Because in the end, true understanding is more important than the mathematical equations. I defy anyone to present the mathematical equations to describe the sound of water dripping from a faucet in a way that an average person can understand. But a person can, in a few sentences using words, give the average person a much better understanding of why he hears water dripping from a faucet.

In Newton's time there were many learned men with advanced models of vortices that could quite accurately predict the movement of the planets. Fortunately for us, Newton did not allow himself to get bogged down in trying to understand the finer details of their theories. In my time there are many books available full of mathematical models to predict things that we do not understand. When one really analyses what their mathematical equations are saying, one soon realizes that they really have no better understanding of the fundamental problems of their subjects than Isaac Newton, Michael Faraday or James Maxwell did. I would encourage anyone interested in furthering the knowledge of science to first and foremost concentrate on what the actual results of particular experiments are. If people can give a lucid explanation in words, it is a good indication that the problem is understood. If, on the other hand, people resort to complicated mathematics without being able to give an explanation in words, then my feeling is that generally the more complicated the mathematics is, the less is really known about the subject.

One of the biggest problems that I have had over the years has been in understanding Maxwell's writing on electricity, magnetism and electromagnetic radiation. The reason for this is that Maxwell presented his work in very mathematical terms that I find difficult to follow. While this may have been necessary to derive what he did, I cannot help but feel that Maxwell's work could be explained far better if words were used instead of mathematical equations.

As I mentioned in the introduction to this work, my feeling is that the final solution to the problems of electricity and magnetism will be found by a person who originally started developing his ideas in his youth. I feel that it is very important that people who understand Maxwell's work get together and translate it in such a way that a young person of twelve or thirteen can read it by himself, without help from others, and understand Maxwell's fundamental ideas without having to get into all the mathematical details that can take decades to learn.

Newton's work, both his Principia and particularly his Optiks, and Faraday's works are both eminently readable by a youngster with very little knowledge of mathematics apart from a little geometry. The only problem that I had in following Newton's Principia was my lack of understanding of certain terms that Newton used, such as versed sine and sagitta which are not common today. Once I understood those terms I found Newton's work very easy to follow. My feeling is that it is important for the progress of science that Maxwell's work be reduced to the same level of readability. I would urge the scientific community to somehow get this done.

I think that Maxwell might have been facing the same problem that I have, namely trying to understand something while not really knowing the exact mechanics of what is happening. I mentioned earlier that I had spent many years thinking about the best way to present my ideas and I really did. In the end I chose to introduce the symbol ϕ to denote the total pressure that the aether exerts in any direction on matter. The reason for this was that it enabled me to put all the things that I am unsure about into a single concept, or symbol, and to then use that in my mathematical derivation of Newton's law of gravity. I do not know whether gravity moves in a wave like manner, similar to that of light, or whether it is as a result of particles moving vast distances throughout the universe at enormous speeds.

But I do feel confident that which ever way gravity works, that my ideas on energy are correct and that the aether is such that it can move with virtually no resistance through bodies as large as the planets and the sun. And in the end, the main concepts that I wish to convey are my feelings as to what energy is, what the aether is composed of and the vast amount of energy that is present in the aether.

Think of the tremendous amounts of energy released in a super nova and try to understand that this is not energy that has slowly been accumulated over a long period of time but is rather energy that is being supplied on a second by second basis from the aether and that this energy is available in any similar volume of space in the universe as that occupied by the exploding star.

Section 4 - Thoughts on a method of propulsion through the aether.

Introduction:

Shortly after the first men landed on the Moon in 1969, I was struck by the tremendous amount of rocket fuel needed to get a man to the Moon and back. I seem to remember reading at that time that for each pound of material taken to the Moon and returned to Earth, over eleven tons of rocket fuel was required.

I started wondering if it were possible to make use of my concepts of the aether to develop a propulsion system for spacecraft that could be powered by nuclear power. The thoughts that I present here were first developed by me during 1969 when I was 22 years old.

I present them here for what they are worth. My theory could be completely worthless due to my lack of understanding of the mechanism that actually causes magnetism. The reason that I present the theory here is that it is my hope that in trying to develop upon the theory we will either develop a practical method of propulsion through the aether or what we learn, in proving why the theory cannot work, may well lead to a better understanding of the fundamental causes of magnetism.

Theory for a method of propulsion through the aether.

I mentioned in Section 1, under the heading titled Law of transmission of force, the following:

"Suppose a first body is acted upon by an external force originating from, or caused by, a second body and this force requires a finite time to be transmitted between the two bodies. Suppose that at any instant of time, t, the force acting upon the first body had required a time dt to travel from the second body to the first body. Then at this instant of time t the force acting upon the first body will be identical to the force that would result if the force was transmitted instantaneously, i.e. with infinite speed, and the second body was, at time t, at the position and state that it had been at the earlier time t-dt."

I got to thinking about magnetism. While we do not understand what causes magnetism, it very definitely results from some type of transmission of matter through the aether. And, just as in the case of gravity, there must be a finite speed with which magnetism travels through the aether.

Magnetism is quite different from gravity, however, as two magnetized bodies can either be pushed towards each other or pushed away from each other, by the aether, depending upon their orientation with respect to each other.

As I mentioned in my propositions in Section 2, I feel that matter has no properties other than inertia. I also feel that all types of "actions at a distance", such as gravity, magnetism, electrical charges, etc. are not properties of matter itself by are rather caused by some type of interaction between large bodies of slowly moving matter and very much smaller particles of matter in the aether moving at speeds considerably faster than the speed of light.

Even though we do not understand the exact process causing magnetism, if one accepts my propositions then one can still make some intelligent guesses as to certain effects that might be possible to create in the aether.

Let us suppose that we have two identical magnets, A and B, separated from each other by a distance such that it takes a time T for a magnetic effect to travel from A to B or from B to A.

Suppose that we initially have both magnets situated on the X axis, where magnet B has a greater X coordinate than magnet A, and both magnets are aligned in the same direction along the X axis with their north poles pointing in the positive X direction, i.e. the north pole of each magnet has a greater X coordinate than the south pole of that magnet.

As is well known, under these circumstances the two magnets will appear to attract each other, although it is really the aether that is pushing the magnets towards each other.

Although we do not understand what is really happening, it is reasonable to suppose that the magnetic force of magnet B upon magnet A is somehow being caused by some disturbance of the aether around magnet A caused by magnet B. Similarly, the magnetic force upon magnet B is somehow being caused by a similar, but quite independent, disturbance of the aether around magnet B that is caused by magnet A.

In the situation where magnets A and B are stationary over a long period of time, it is well known that the magnetic force by magnet A upon magnet B is equivalent to the magnetic force by magnet B upon magnet A but acting in the opposite direction. It is reasonable to suppose, as was shown in the case of gravity between two bodies, that this is due to the fact that the relative positions of the two magnets with respect to the aether does not change. Just as was illustrated in the case of gravity, however, it would be a mistake to assume that Newton's third law could be used to predict that the two forces would always be identical or even acting in the same directions.

Suppose that for a long period of time the two magnets, A and B have remained at rest in the aether. Magnet A would experience a magnetic force pushing it towards magnet B and magnet B would experience a magnetic force pushing it towards magnet A.

Now suppose that at time t=0, we could suddenly and instantaneously rotate magnet A through 180 degrees so that the south pole of magnet A would now have a greater X coordinate than the north pole of magnet A. Suppose further that we rotated the magnet so that the midpoint between the poles of magnet A remained the same distance to the midpoint between the poles of magnet B as before.

Let us now think about what would happen to magnets A and B.

Once magnet A is rotated by 180 degrees at time t=0, then whatever property it is in the aether, caused by magnet B, that originally caused magnet A to be pushed towards magnet B will now cause magnet A to be pushed away from magnet B. So during time t=0 until t=T magnet A will be pushed away from magnet B. During the time period t=0 until t=T, however, magnet B would not be aware of the fact that the poles of magnet A have been switched around due to the fact that, in our example, it takes a time T for the magnetic effect caused by magnet A to travel from magnet A to magnet B. Hence during this period t=0 until t=T magnet B will be pushed towards magnet A as it was previously.

Hence during the time period t=0 until time t=T we have an unusual effect in that magnet A is being pushed away from magnet B while at the same time magnet B is being pushed towards magnet A.

Now at time t=T, magnet B will become aware that the poles of magnet A were switched around at time t=0. At time t=T magnet B would, therefore, start to be pushed away from magnet A by the aether.

Let us suppose that we now, at time t=T, suddenly and instantaneously rotate magnet B through 180 degrees so that the south pole of magnet B would have a greater X coordinate than it's north pole. At this point, therefore, magnet B would "become aware" that it's poles were now lined up in the same direction as the poles of magnet A appear to be. Magnet B would then once again start to be pushed towards magnet A.

Now let's think about what happens to magnet A during the time period t=T until t=2T. During this period, magnet A is being affected by magnet B as it was at a time T earlier, namely during the period T=0 until t=T. So during this period t=T until t=2T we once again have the unusual effect that magnet A is being pushed away from magnet B whereas magnet B is being pushed towards magnet A.

Now at time t=2T, magnet A will suddenly become aware of the fact that the poles of magnet B were switched around at time t=T. Magnet A would hence immediately start being pushed towards magnet B.

Now let us suppose that at time t=2T we once again suddenly and instantaneously rotate the poles of magnet A through 180 degrees again so that the north pole of magnet A would now have a greater X coordinate than it's south pole. So once again, magnet A would start to be pushed away from magnet B.

Now lets think about what happens to magnet B during the time period t=2T until t=3T. During this period magnet B is being affected by magnet A as it was at a time T earlier, namely between time t=T until t=2T. Hence during this period t=2T until t=3T, magnet B will be being pushed towards magnet A. So during this period t=2T until t=3T we once again have the unusual effect that magnet A is being pushed away from magnet B whereas magnet B is being pushed towards magnet A.

Now at time t=3T, magnet B will become aware of the face that the poles of magnet A were switched around at time t=2T. Hence, at time t=3T, magnet B will start to be pushed away from magnet A.

Now suppose that at time t=3T, we once again instantaneously rotate magnet B trough 180 degrees so that the north pole of magnet B had a greater X coordinate than it's south pole. So once again, magnet B would start to be pushed towards magnet A.

Now lets think about what happens to magnet A during the time period t=3T until t=4T. During this period magnet A is being affected by magnet B as it was at a time T earlier, namely between time t=2T until t=3T. Hence during this period t=3T until t=4T, magnet A will be being pushed away from magnet B. So during this period t=3T until t=4T we once again have the unusual effect that magnet A is being pushed away from magnet B whereas magnet B is being pushed towards magnet A.

Now at time t=4T, magnet A will suddenly become aware of the fact that the poles of magnet B were switched around at time t=3T. Magnet A would hence immediately start being pushed towards magnet B.

Now let us suppose that at time t=4T we once again suddenly and instantaneously rotate the poles of magnet A through 180 degrees again so that the south pole of magnet A would now have a greater X coordinate than it's north pole. So once again, magnet A would start to be pushed away from magnet B.

The table, Section 4 - Table 1, on the next page illustrates what has been discussed above:

Time Period t	Actual Direction of Magnet A	Perceived Direction of Magnet B by Magnet A	Direction of Force acting on Magnet A	Actual Direction of Magnet B	Perceived Direction of Magnet A by Magnet B	Direction of Force acting on Magnet B
Prior to 0	S-N	S-N	Towards B	S-N	S-N	Towards A
0 to T	N-S	S-N	Away From B	S-N	S-N	Towards A
T to 2T	N-S	S-N	Away From B	N-S	N-S	Towards A
2T to 3T	S-N	N-S	Away From B	N-S	N-S	Towards A
3T to 4T	S-N	N-S	Away From B	S-N	S-N	Towards A
4T to 5T	N-S	S-N	Away From B	S-N	S-N	Towards A
5T to 6T	N-S	S-N	Away From B	N-S	N-S	Towards A
6T to 7T	S-N	N-S	Away From B	N-S	N-S	Towards A
7T to 8T	S-N	N-S	Away From B	S-N	S-N	Towards A
8T to 9T	N-S	S-N	Away From B	S-N	S-N	Towards A

Section 4 - Table 1

where:

S-N indicates a magnet aligned along the X axis with the north pole having a greater X coordinate than the south pole.

N-S indicates a magnet aligned along the X axis with the south pole having a greater X coordinate than the north pole.

If one examines all of the above, one will see that during the entire period t=0 until t=4T, we had the unusual situation where magnet A was being pushed away from magnet B while at the same time magnet B was being pushed towards magnet A.

Furthermore, from time t=4T until t=5T the conditions of magnets A and B are identical to what they were between period t=0 until t=T. Hence by repeating what was done during period t=0 until t=4T during the period t=4T until t=8T we would arrive at the situation that for the entire period t=0 until t=8T magnet A was continually being pushed away from magnet B while during the same period magnet B was continually being pushed towards magnet A.

And obviously, by repeating this cycle of period 4T ad infinitum, we could arrive at the fascinating situation that magnet A would be continuously be pushed away from magnet B whereas magnet B would be continuously being pushed towards magnet A. Hence both magnets would be continually pushed in the same direction.

Let us now imagine that we could mount both magnets, A and B, in some type of vessel such that any force acting upon the magnets would be transmitted to the vessel. Under such circumstances, we would have a vessel that would experience a continual force pushing it in the direction from magnet B towards magnet A. And due to the fact that it is actually the aether that is pushing the vessel, and the fact that the aether exists everywhere in space, there is no reason why the vessel could not be used as a vessel to travel between the planets and ultimately between stars.

This method of generating propulsion through the aether by continuously switching the direction of the poles of two magnets reminds me in some ways of the way in which a row boat is propelled through the water by continually moving oars backwards and forwards. I like to think of this method of switching the poles of magnets around continually as being a method of "rowing" through the aether.

Thoughts on symmetry between the two magnets.

If a system such as the one described above ever could be built, then the one thing that becomes immediately obvious is that there would be a lack of symmetry about the system which is quite unlike anything currently known. In all types of gravitational effects, every piece of atomic matter appears to be pushed towards every other piece of atomic matter. In all effects currently known in electrostatics or magnetism, two bodies are either mutually pushed towards each other or are mutually pushed away from each other. There is a certain symmetry to this which has, in the past, caused people to assume that Newton's third law of motion is applicable between the two bodies.

In the case of the two magnets described above, the first magnet is continually being pushed away from the second whereas the second is continually being pushed towards the first. This lack of symmetry initially bothered me and I felt that something must be wrong with my concept.

When one really thinks about it, however, this lack of symmetry is brought about by a lack of symmetry in the way in which the poles of the magnets are turned around. If one looks at the table above, it is readily apparent that the switching time of magnet A is one quarter of the full cycle of 4T ahead of that of magnet B. This asymmetry is the ultimate reason for the lack of symmetry between magnet A and magnet B. In fact, if one simply altered the asymmetry of the switching between the two magnets such that magnet B was one quarter of the full cycle of 4T ahead of magnet A, then the direction of motion of the magnets would be reversed so that magnet B was continually being pushed away from magnet A and magnet A was continually being pushed towards magnet B. If one kept the full cycle of 4T the same but changed the switching that the two magnets were actually either in phase with each other or half of the full cycle of 4T out of step with each other then once again there would be symmetry in that at any time both magnets would either be pushed towards each other or away from each other twice each full cycle of 4T. The tables below, Section 4 - Table 2 and Section 4 - Table 3, indicate this.

Table showing effect of switching magnets A and B in phase with each other.

Time Period t	Actual Direction of Magnet A	Perceived Direction of Magnet B by Magnet A	Direction of Force acting on Magnet A	Actual Direction of Magnet B	Perceived Direction of Magnet A by Magnet B	Direction of Force acting on Magnet B
Prior to 0	S-N	S-N	Towards B	S-N	S-N	Towards A
0 to T	N-S	S-N	Away From B	N-S	S-N	Away From A
T to 2T	N-S	N-S	Towards B	N-S	N-S	Towards A
2T to 3T	S-N	N-S	Away From B	S-N	N-S	Away From A
3T to 4T	S-N	S-N	Towards B	S-N	S-N	Towards A
4T to 5T	N-S	S-N	Away From B	N-S	S-N	Away From A
5T to 6T	N-S	N-S	Towards B	N-S	N-S	Towards A
6T to 7T	S-N	N-S	Away From B	S-N	N-S	Away From A
7T to 8T	S-N	S-N	Towards B	S-N	S-N	Towards A
8T to 9T	N-S	S-N	Away From B	N-S	S-N	Away From A

Section 4 - Table 2

Table showing effect of switching magnets A and B half a cycle out of phase with each other.

Time Period	Actual Direction of Magnet A	Perceived Direction of Magnet B by Magnet A	Direction of Force acting on Magnet A	Actual Direction of Magnet B	Perceived Direction of Magnet A by Magnet B	Direction of Force acting on Magnet B
Prior to 0	S-N	S-N	Away From B	N-S	N-S	Away From A
0 to T	N-S	N-S	Towards B	S-N	S-N	Towards A
T to 2T	N-S	S-N	Away From B	S-N	N-S	Away From A
2T to 3T	S-N	S-N	Towards B	N-S	N-S	Towards A
3T to 4T	S-N	N-S	Away From B	N-S	S-N	Away From A
4T to 5T	N-S	N-S	Towards B	S-N	S-N	Towards A
5T to 6T	N-S	S-N	Away From B	S-N	N-S	Away From A
6T to 7T	S-N	S-N	Towards B	N-S	N-S	Towards A
7T to 8T	S-N	N-S	Away From B	N-S	S-N	Away From A
8T to 9T	N-S	N-S	Towards B	S-N	S-N	Towards A

Section 4 - Table 3

The interesting thing shown in the two tables above is that in both cases the two magnets would either both be pushed towards each other or away from each other at any instant of time. The time taken to alternate between pushed towards each other or away from each other would be T, or half the time taken to switch the poles of the magnets around. In addition, if one were to sum up the total forces acting on each of the magnets over a long period of time, such as a second, the net force acting upon each magnet would be zero. A casual observer, not aware of the rapid speed at which the magnets were being switched, would fail to detect any observable magnetic effect in either of the magnets.

Thoughts on solutions to practical problems involved in building such a vessel.

When I first thought of this method of propulsion in 1969 it seemed to me impossible at that time that such a system of magnets could ever be built. At that time, the computer that I worked on was built with circuit boards containing individual transistors and electrical switching times were impossibly slow.

Since that time, however, electrical switching time have increased to the gigahertz level and expertise at creating integrated circuits has increased enormously. In addition, chemical technology in creating extremely small magnetic particles on the surfaces of disks drives has improved enormously. It does not seem at all far fetched to me now that technology could one day progress to the stage whereby it might be possible to make a practical device that can "row" through the aether.

The first problem in making such a device is obviously is one of being able to switch magnets around instantaneously. Obviously one is going to run into problems of mutual induction and self induction. The closer that one can approach this ideal of instantaneous switching, the more efficient the mechanism would be.

If, however, one merely used alternating currents in coils where the frequency of the current was 4T between the two coils, and the current in the leading magnet was a quarter of a cycle ahead of the trailing coil, then one could still achieve the same effect but with reduced efficiency.

The second problem is one of building coils that can be switched rapidly enough within the time that it takes magnetism to travel between the two coils. If one is trying to create switchable magnets using electric coils, then obviously the total distance of the winding in the coil has to be very much less than the distance between the coils pushing and pulling each other in order to allow current to be established within the coil itself. This indicates to me that one will need, instead of a single magnet, to create flat, extremely thin arrays of coils, all switched in parallel with each other at exactly the same time. If one could construct two such sheets of arrays of magnets and arrange them at exactly the correct distance apart, then one might get some useful effect.

If one considers very large arrays of extremely small coils on parallel plates, then there will obviously need to be a sophisticated system of triggering the switching of the coils due to the fact that it will take time for the triggering signal to travel across the arrays. The times required for this are, however, predictable and by changing the point during the triggering signal at which any particular coil is triggered to take into account the delay involved for the signal itself to pass between coils, it should still be possible to get all arrays to switch at the correct times from a single signal source.

Final thoughts on the proposed method of rowing through the aether.

If such a system of electro magnets could ever be built, then one would have the means to very easily leave the Earth and more importantly the Sun. Electrical energy would still have to be supplied to the system to continually change the direction of the magnetic fields and to supply the power that would be needed to lift any such vessel off the Earth and away from the Sun. The electrical energy could, however, be generated from nuclear fission or fusion. Due to the fact that the propulsion system would use the aether itself, moving at very high velocities, to push the vessel forward, one would not need to eject the tremendous amounts of matter at low speed that is currently done with chemical rockets. This would make it possible for a nuclear powered space vessel to leave the Earth, visit other planets and moons in the solar system and return to Earth and land on it without having lost much in the way of total matter except that consumed in the nuclear reaction used to generate the electricity.

Over the years I have often wondered what it will be like when people can quickly and easily fly between the planets and what people will be thinking when they are on such ships. Early men probably started dreaming of being able to fly ever since they first watched birds fly. Early Greek mythology as well as Arabian tales contain various stories of people flying on various things such magic carpets and wings made of bird feathers

Leonardo da Vinci definitely imagined people being able to fly and he probably wondered what fun it would be to be able to fly through the air. He might even have been prepared to give almost everything that he had to be able to fly on a plane that we take for granted today. He might also have wondered what people aboard planes would think when traveling through the air in much the same way that we wonder about what it will be like to be able to travel between the stars. He might even have felt some regret, in his later years, as he realized that he would never live to see men fly through the air.

I do feel some regret that I will never live to see men fly between the stars. But I feel very content that I have lived at the time that I have. I feel that in the long term, many of the most important scientific experiments will be conducted in deep space, far from any stars. I do not feel, however, that I would be prepared to give all that much to be able to travel on one of those great space ships of the future. After having observed many people flying aboard planes, my feeling is that when our descendants are traveling between the planets and the stars in wonderful space vessels, the main concern for most of them will probably be how much longer it will be before they arrive at their destination and what will be served for their next meal. And when men populate the stars, many of them may well dream of being able, in their lifetimes, to visit our own Earth for a few days and to be able to savor everything on Earth that we unfortunately take for granted today.

It is my firm belief that when our descendants in ten thousand years look back at us they will not remember us as being the first ones to reach the Moon or for all our other technical achievements that have taken place in the past few hundred years. I feel that they will rather look back upon us as the ones who, in our greed and ignorance about what is really important, caused and allowed so many species of plants and animals on Earth to become extinct. And in doing so risked causing mankind itself to become extinct.

Section 5 - Thoughts on angles.

Introduction.

One of the classical problems in geometry was to trisect an angle using a compass and a ruler. It has been proved mathematically that this cannot be done. Whilst an exact solution is not possible, I thought of a practical way to trisect any angle, using a compass and a ruler, to any desired degree of accuracy. I thought of this one morning while in the shower looking at the patterns formed by the tiles on the wall. Since then, I have extended the original concept to enable one to construct any angle with the use of a computer, a compass and a ruler. All geometric problems requiring the use of a compass and ruler also implicitly require the use of a human brain as well. I do not see any reason, therefore, why we should not extend our classical concepts somewhat to allow the use of a computer as well as a compass and a ruler in solving geometric problems.

Once I had accepted the logic of allowing one to use a computer in geometry, I later realized that by using a simplified system to measure angles, it would be possible for any twelve year old beginner in geometry to construct any specified angle by simply using a computer, a compass and a ruler.

We currently have two methods of measuring angles. One method uses a system which, as far as I can tell, dates back to Sumerian times in which a circle is divided into 360 degrees. The reason for choosing 360 degrees in a circle is that it makes it fairly simple for one to divide a circle in to a large number of different equal parts. 360 is divisible by 22 different numbers, namely 2, 3, 4, 5, 6, 8, 9, 10, 12, 15, 18, 20, 24, 30, 36, 40, 45, 60, 72, 90, 120 and 180. This system has been in use for thousands of years and it obviously satisfies the need to be able to divide a circle into a large number of different equal portions.

The second system of measuring angles that we have is in radians. This system is relied upon in all of our mathematical series used to calculate sines, cosines, tangents etc.

One of the problems that we will encounter is that as we look further and further into the universe, we will need to divide angles into smaller and smaller parts. The system that I will propose will cater for this, both in terms of being able to accurately generate an angle to any degree of precision as well as in terms of measuring an angle to any degree of precision.

An algorithm to trisect any angle to any degree of precision using a compass and a ruler.

Method:

Suppose that one has an angle formed by the intersection of two lines. Two different lines trisect this angle at positions one third and two thirds of the angle between the two original lines. Choose which of one of the two trisectors one wishes to find. Number the original lines as 1 and 2 where line 1 is the furthest line from the trisector that one wishes to find.

Bisect the original angle between lines 1 and 2. Number this bisector as line 3. Now bisect the angle between lines 2 and 3 and number this bisector as line 4. Now bisect the angle between lines 3 and 4 and number this new bisector as line 5. Keep on repeating this process of bisecting the angle between the two highest numbered lines and numbering the new line formed, i.e. the bisector, as one higher than the previously highest numbered line. The trisector sought will always lie between the two highest numbered lines.

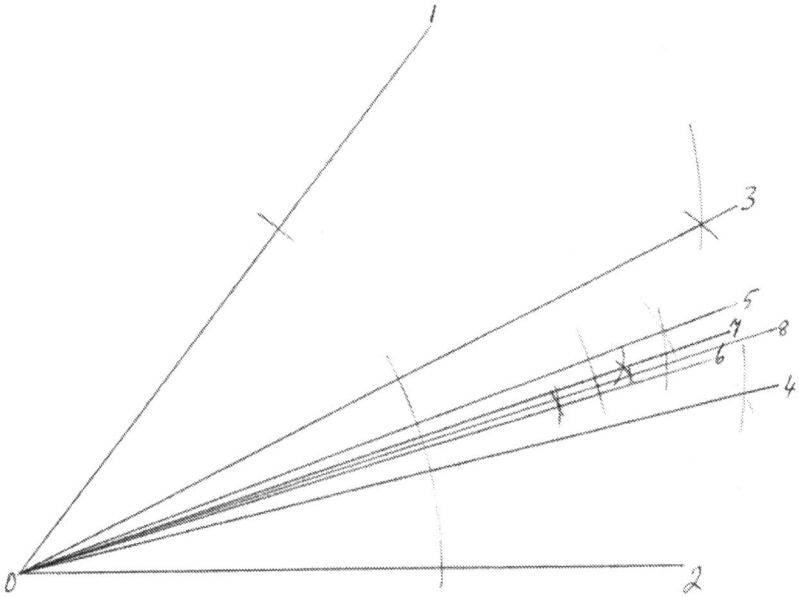

Section 5 - Figure 1 - Trisecting an angle.

Proof:

Let S be the sum of the series 1/2 - 1/4 + 1/8 - 1/16 + 1/32 - 1/64 + 1/128 − 1/256 + ...

So:

$S =$ +1/2 - 1/4 + 1/8 - 1/16 + 1/32 - 1/64 + 1/128 − ......

Multiplying both sides by 2 gives:

$2S = 1$ - 1/2 + 1/4 - 1/8 + 1/16 - 1/32 + 1/64 - 1/128 + ...

Adding both lines gives:

$3S = 1$

Hence

$S = 1/3$

Therefore

$1/3 =$ +1/2 - 1/4 + 1/8 - 1/16 + 1/32 - 1/64 + 1/128 − ......

Each term of the series for 1/3 is thus minus one half of the previous term.

According to the above equation, to get one third of an angle, first start with half of the angle. In the figure 5.1 above, line 3 represents half of the angle between lines 1 and 2. Then subtract one half of this angle (-1/4). Line 4 represents subtracting one half of the angle between 2 and 3. Then add one half of the value of the angle between 3 and 4 (+1/8). Line 5 represents adding half of the angle between lines 3 and 4. Then subtract one half of the angle between angle 4 and 5 (-1/16). Line 6 represents subtracting one half of the angle between lines 4 and 5. Then add one half of the angle between line 5 and 6 (+1/32). Line 7 represents one half of the angle between lines 5 and 6. And so on ad infinitum. In the figure above, the trisector will lie one third of the way between lines 7 and 8. After 12 bisections, one will have determined the position of the trisector to an accuracy of one part in 4096 which is probably far more accurate than it is possible to construct with a compass and ruler on a piece of paper.

Method to construct any fraction of any angle using a computer, compass and ruler.

When one thinks about what was done above in trisecting an angle, one will realize that we were alternatively adding or subtracting half of whatever angle we had just created as the need arose. It just happens that 1/3 happens to be formed by an extremely easy and interesting series.

If we expressed 1/3 and 2/3 in binary notation we would have:
1/3 = 0.0101010101010101010101010101010101 ..... B
2/3 = 0.1010101010101010101010101010101010 ..... B

We used our alternating pattern of ones and zeros to create the simple algorithm shown above for trisecting an angle.

Now any fraction of any angle can be expressed in binary notation.

So
1/5 = 0.001100110011001100110011001100110011001100..B
5/13 = 0.0110001001110110001001110110001001110110...B
etc.

Therefore, if we can extend our algorithm used above in trisecting an angle to allow us to simply add or subtract half of the previous angle, as the need arises, then we can simply and accurately construct any fraction of any angle.

The first thing that we need is a computer program that can print out the binary representation of any fraction. I have enclosed, at the end of this section, a listing of a simple function that I wrote as a Visual Basic macro for an Excel spreadsheet that will express any fractional number in binary notation. The routine is called double_to_binary. I have also enclosed a listing of a slightly more complicated Excel function to express any fractional number in hexadecimal. This routine is called double_to_hex. I do not wish to waste any time explaining how the programs work other than to say that I am sure that any competent programmer can verify that the programs are accurate to as many bits of precision as there are in the mantissa of a double precision floating point number. The accompanying routines can easily be copied by the average user of Excel into a spreadsheet macro and used directly without the need to know much about programming at all. These functions might already be available in some scientific calculators and very efficient forms of the functions could easily be incorporated into new designs of calculators if the need was there.

Suppose that we have any angle formed by two lines, numbered 1 and 2, intersecting at point O, as in the Figure 5.2 shown below. Let us bisect the angle between lines 1 and 2 and number the line bisecting the angle as line 3. Suppose that we use binary notation to refer to the angle formed by the two lines 1 and 2. Let us, as is customary, assume that an angle increases as we move anti-clockwise from line 1 to line 2. On this basis, line 1 would represent 0.0B of the total angle between lines 1 and 2 and line 2 would represent 1.0B of the total angle. Line 3 would represent one half of the angle, or 0.1B of the angle, between lines 1 and 2.

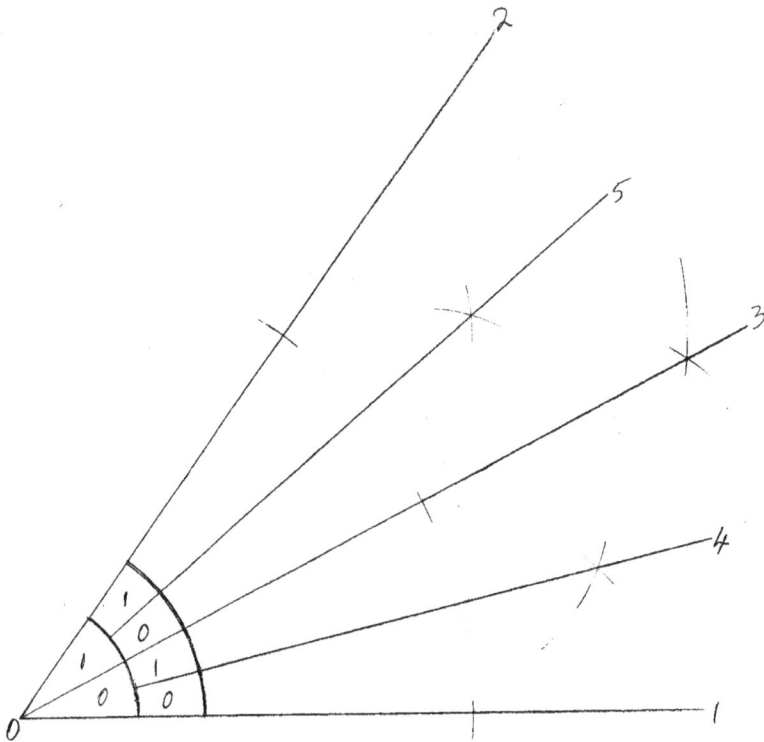

Section 5 - Figure 2.

Hence if one were to draw any line through point O between lines 3 and 1, the angle formed between that line and line 1 would be in the range of 0.0B to 0.0111111111...B of the original angle between lines 1 and 2. Similarly, if any line were drawn through point O between lines 3 and 2, then the angle formed between that line and line 1 would be in the range 0.1B to 0.1111111111....B of the angle between lines 1 and 2. Let us number the angle between line 1 and 3 as 0 and number the angle between line 3 and 2 as 1. These two numbers, 0 and 1, would represent the first binary digits of the value of the fraction of the angle formed by any line drawn through point O, between line 1 and 2, and line 1 with respect to the angle between lines 1 and 2.

Now let us bisect the angles 0 and 1 with two new lines, 4 and 5. For each angle bisected, number the two new angles formed as 0 and 1 as shown in figure 5.2 above. Using the same logic as above, the angle between line 1 and 4 represents one half, or 0.1B, of the angle between lines 1 and 3. But the line between angle 1 and 3 is 0.1B of the angle between line 1 and 2. Hence the angle between lines 4 and 1 is 0.1B times 0.1B of the angle between line 1 and 2. This value is 0.01B, or one quarter. Similarly the angle between lines 5 and 3 is 0.01B of the angle between lines between angle 1 and 2. Hence the angle between line 5 and line 1 is simply 0.10B + 0.01B, or 0.11B, of the angle between lines 1 and 2. Hence as can be seen, any line drawn between line 2 and 5 will form an angle with line 1 that will fall in the range of 0.11B to 0.111111111....B of the angle between line 1 and 2. Similarly, any line drawn between line 5 and 3 will form an angle with line 1 that will fall in the range of 0.10 to 0.101111111111....B of the angle between lines 1 and 2. Similarly, any line drawn through point O between line 3 and line 4 will form an angle with line 1 that will fall in the range of 0.01B to 0.011111111.....B of the angle between lines 1 and 2. And finally, any line drawn through point O between lines 1 and 4 will form an angle with line 1 that falls in the range of 0.0B and 0.00111111111111B of the angle between lines 1 and 2.

Looking at figure 5.2, it will be seen that if we combine the notations of the angles created, from the point O outwards, in a near to far manner from point O, then the two digits formed give the first two digits of the binary fraction any line drawn in that range. Hence, reading the angles outwards from point O, any line drawn between lines 5 and 3 would form an angle with line 1 which, when expressed as a binary fraction of the angle between lines 1 and 2, would have to begin with the binary digits 0.10....

Figure 5.3 shows this concept extended to 4 levels of bisections. Looking at the angle between lines A and B, and the numbers used to number these lines, it can easily be seen that any line drawn through point O between lines A and B, would lie in the range 0.0110B to 0.01101111111...B of the angle between lines 1 and 2.

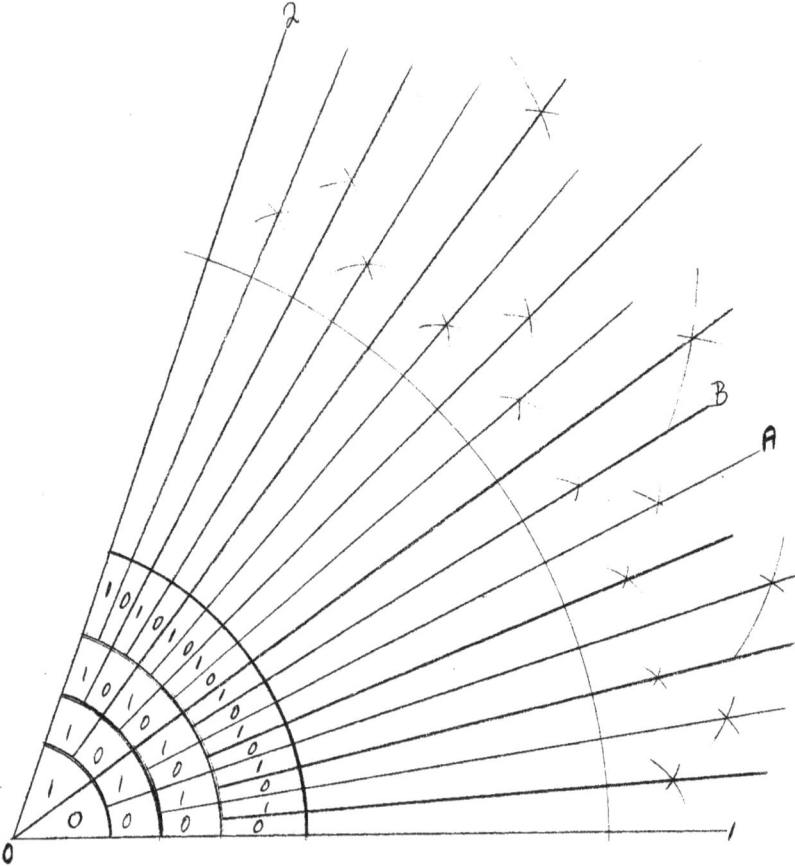

Section 5 - Figure 3

Hence we now have a simple method to create any fraction of any angle by means of a computer, compass and ruler. If we simply use a computer to print out the binary representation of the angle desired, one can then use a compass and ruler to create the angle by simply reading the binary value of the fraction desired as an instruction list of the order in which angles must be bisected.

Example to obtain 5/13 of an angle using a computer, compass and ruler.

Using the program supplied, one finds double_to_binary(5/13) = 0.011000100111011000100111011000100111011000100...B

Construct the angle to be divided by drawing two lines, numbered as 1 and 2, intersecting at point O as shown in figure 5.4. Bisect this angle forming two new angles, numbered 0 and 1. Since the first binary digit of the fraction required, .0110...., is a 0, bisect the angle 0 forming two new angles 0 and 1. Since the next binary digit in the fraction is a 1, bisect the last angle 1 formed forming two new angles 0 and 1. Since the next digit is a 1, bisect the last angle 1 formed forming two new angles 0 and 1. Since the next digit is a 0, bisect the last angle 0 formed forming two new angles 0 and 1. Since the next digit is a zero, bisect the last angle 0 forming two new angles 0 and 1. Since the next digit is a zero (we are now up to 0.011000), bisect the last angle 0 forming two new angles 0 and 1. Since the next digit is a 1 (0.0110001) bisect the last angle 1 created forming two new angles 0 and 1. Continue in this way to any degree of precision required. In figure 5.4, the line sought lies between lines 8 and 9.

By using the method described in this example, we have a simple method to obtain any fraction of any angle to any degree of precision required by simply using a computer to display the value of the fraction required in binary, to at least the degree of precision required, and then using a compass and ruler to repeatedly bisect angles depending upon the binary digits in the desired fraction.

Section 5 - Figure 4 - Obtaining 5/13 of any angle.

Thoughts on yet another way to measure and describe angles.

The method that I have outlined above makes it very easy, with the aid of a computer, to easily obtain any fraction of any angle and to easily divide a circle into any number of equal parts.

One of the problems with the 360 degree system is that in many situations a degree is a very large unit of measure. In astronomy, as well as in navigation, a degree is routinely divided into smaller units of minutes and seconds of an arc. As we look further and further into the universe, it is obvious that we will continually need to split up whatever unit we use into smaller units. Using minutes and seconds of arcs is cumbersome, as factors of 60 are introduced. Once one gets to minutes and seconds of an arc, I find it extremely difficult to visualize these angles and I doubt that any child could easily draw an angle of, for example, 23 degrees, 12 minutes and 46 seconds of an arc. The other system that we have of working with angles, namely in radians, has the problem that until now no simple means existed to accurately draw an angle of one radian.

I feel that, in situations where one needs to accurately measure or construct angles, a very much simpler method of describing angles would simply be to assign a new unit of angular measurement such that one unit represents one complete rotation of a line about a point, or 360 degrees. Any angle can then be expressed as fraction of this unit. If the angle is expressed in binary notation, then the angle could easily be drawn by a twelve year old child with the aid of a compass and a ruler.

Example to construct an angle of 1 radian using a computer, compass and a ruler.

Given that one of our new units of angular measurements represents a complete revolution, or 360 degrees, there are $2.\pi$ radians in our new angular unit. One radian is, therefore, $1/(2.\pi)$ of our new unit of angular measurement.

Using the program that I have supplied in Excel, one finds that
double_to_binary($1 / (2 * PI())$)
$= 0.0010\ 1000\ 1011\ 1110\ 0110\ 0000\ 1101\ 1011\ 1001\ ....B$

and double_to_hex(1 / (2 * PI()))= 0.28BE 60DB 9391 04...h

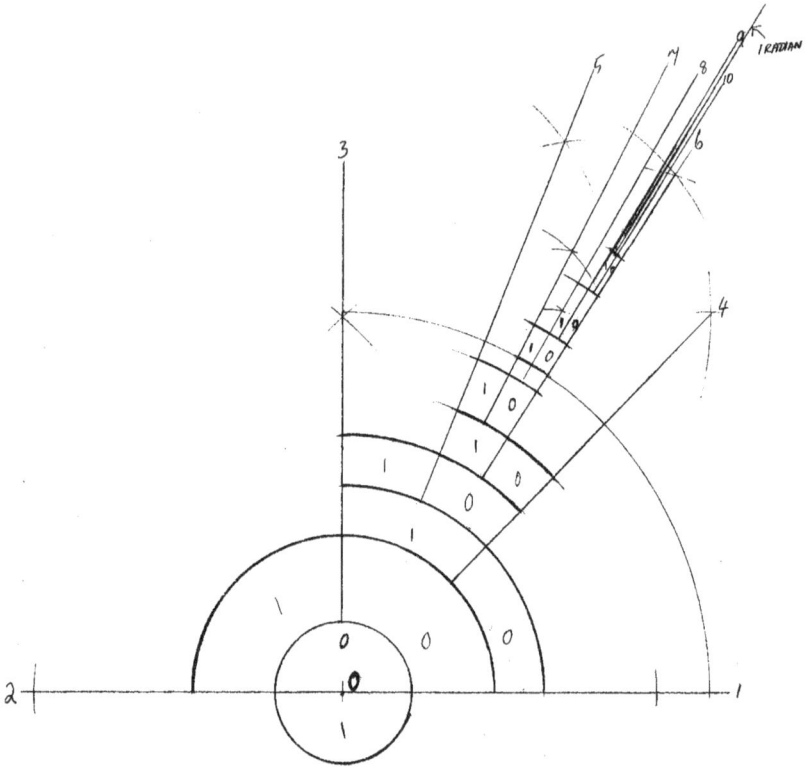

Section 5 - Figure 5 - Constructing an angle of 1 Radian
(0.0010 1000 1011 1110 0110 0000...B).

Draws a straight line and mark a point on it as point O. Number the portion of the line to the right of point O as 1. Number the portion of the line to the left of point O as line 2. Line 2 would thus represent bisecting the total angle around point O formed by rotating line 1 through one complete revolution about point O back to its original position. Hence mark the top angle formed between line 2 and 1 as 0 and the bottom, or second angle between line 2 and line 1 rotated through one revolution, as angle 1. These two angles 0, and 1, thus represent the starting angles used in the algorithm mentioned above to divide any angle where, in this case the angle is one complete rotation. Using these starting angles and the value of the fraction required as 0.001010001011... etc., a twelve year old can easily construct an angle with the value of 1 radian as shown in figure 5.5 above. The line representing 1 radian lies between lines 9 and 10, very nearly on the bisector of line 9 and 10.

Further thoughts on ways of expressing angles.

Once one starts to express angles simply as a fraction of a unit, the unit being one complete revolution or 360 degrees, one has a very simple way of extending the precision to which an angle is expressed. If the angle is expressed in binary units as a binary fraction, one can simply extend the number of fractional binary digits to any degree of precision required. All the complications of dealing in degrees, minutes, seconds and fractions of seconds are eliminated.

If one thinks about it, once any angle is simply described a fraction of one unit, the unit being 360 degrees, then it does not really matter what base is used, binary, decimal or hexadecimal. Computers can very easily handle any base.

For people interested in constructing angles, using a compass and ruler, or any other mechanical system that is based on the concept of repeatedly dividing an angle, then my personal feeling is that the hexadecimal system has some advantages over the binary system in that it is, at least for me, simpler to remember various numbers expressed in hexadecimal as opposed to binary.

I find it quite simple to remember that a radian has a hexadecimal value of 0.28BE60...H. If a young student merely has to remembers that the first three hexadecimal digits of the value of one radian as 0.28BH then, knowing how to expand each hexadecimal digit into binary, the student could easily, with a compass and a ruler, construct an angle with a value of 1 radian to an accuracy of within one part in four thousand, or about one tenth of a degree.

Thoughts on measuring angles.

The methods outlined above gives one the ability to very simply construct an angle with any value that one desires. The reverse is equally true in that it is just as easy to measure any given angle, particularly if our unit of measurement is one rotation, or 360 degrees. By starting with a straight line being one of the sides of a given angle, and repeatedly dividing the unit angle by 2 and choosing which of the two halves of each resultant angle to use based upon where the other side of the angle is, one can, by noting the progression of the subdivisions, easily measure the value of any angle in binary and subsequently convert it to any other base desired.

Thoughts on measuring longitudes.

The British people established the standard of measuring longitude in degrees. They did it in a slightly unusual way, however, in that a line on the floor of the Greenwich observatory was chosen as zero degrees longitude. In addition, they then added a typically British complication to the problem by measuring longitude to the East and West of Greenwich in terms of degrees either to the East or West of Greenwich. So, for example, New York, USA, has a longitude of 75 degrees west of Greenwich and Mumbai (Bombay), India, has a longitude of 75 degrees east of Greenwich. This method had it's advantages at the time when England and Europe were the centers of scientific development. One advantage of choosing Greenwich as the zero degree of longitude is that the international date line, which is at 180 degrees east or west of Greenwich, falls mainly across the Pacific Ocean where there are few land masses.

The problem with the system, however, is that whereas it if fairly simple for a person in London, England, to work out the time difference between London and any other place on the Earth, it is a lot more difficult for people near the international date line to easily work out the time difference between themselves and some other place on the other side of the date line. Let's, for instance, imagine that it is 3 p.m. on Monday afternoon in Alaska and suppose that someone in Alaska wishes to work out what time it is in Tokyo. They first have to add their longitude, to the west of Greenwich, to that of Tokyo, which is to the east of Greenwich. They then have to divide that by 360 degrees and then multiply that by 24 to get the number of hours difference. They would then get an answer of 19 which, when subtracted from 24, means in fact 5 hours. Now, since they actually to the east of Tokyo, even though Tokyo is to the East of Greenwich and they are West of Greenwich, then if it is 3 p.m. in Alaska this actually means that it is five hours earlier in Tokyo, or 10 am. However, since Tokyo is on the other side of the international date line, it is really the next day in Tokyo. Hence when it is 3 p.m. on Monday in Alaska it is really 10 am on Tuesday in Tokyo.

Suppose that instead of using the 360 degree system, as is used at present, one were instead to use the system which I proposed earlier to measure angles, namely where one unit of measurement was equal to one rotation, or 360 degrees. In this case, one unit would represent one day. Suppose further that we chose the longitude of the international date line, 180° East or West of Greenwich, to represent 0 or 1 unit of longitude. In that case, London England would be at longitude 0.5 in decimal or 0.10 in binary or 0.80 in hexadecimal. Suppose further, that we used as the direction of rotation of the angle the direction that the sun appears to move, i.e. from East to West. So going from 0 unit of longitude at the international date line, longitude would gradually increase until at about 0.25 decimal, or 0.01B or 0.40h, one would be over Asia. At 0.5 decimal, or 0.10B or 0.80h one would be at the longitude of London. As one continues, one would reach

0.75 decimal, or 0.11B or 0.C0h at the longitude of Chicago in America.

If one were at any place on Earth, to get the time difference between oneself and any other place one would simply subtract the longitude of the other place, in fractions of a rotation, from one's own longitude. It the result is positive, it means that the time of day in that place is later in the day than the time of day in the place that you are at. If the result is negative, it means that the time of day in the other place is earlier in the day than the time of day that you are in. If one simply multiplies the result of the subtraction by twenty four, one will get the number of hours in the day that the other place is either ahead of or behind you by. One then simply adds the signed number of hours to one's current time. If adding and the time moves past 24 hours, then the other place is a day ahead of one calendar wise. If adding and the result is negative, then the other place is a day behind one calendar wise.

The major advantage of the system proposed is that the same simple rules for calculating time differences will work anywhere on Earth. All the difficulties involved in having to worry about whether a place is to the East or West of Greenwich will be eliminated.

Finally, if one still wishes to keep the 360 degree system, but base it on the same basis as the angles measured above, then 0 degrees would be at the international date line. Mumbai (Bombay) would be at 105 degrees. London would be at 180 degrees. Chicago would be at 270 degrees. The difference now, however, would be that the same rules for calculating time differences could be used anywhere on Earth.

' **Visual Basic Macros for Excel to convert double precision floating point number to binary**
' **or hexadecimal representation**
‘
' **The following two routines may be copied by anyone and used for any purpose.**
'

```vb
Function double_to_binary(input_value As Double) As String
'   Author:  Ian Atkinson
'   Date:  February 28, 2006
'   This function converts a floating point double precision number into a
'   character string containing the binary representation of the floating point number.
'   Limitations:
'   Numbers greater or equal to 4294967296 not allowed
'   Number of characters in result dependent on number of bits in the mantissa
'   of the computer representation of a double precision floating point number.
'   Last output digit is rounded result of last digit and digits truncated.

    Dim result_string As String    ' result_string contains the binary chars so far in the result
    Dim power_val As Double         ' This value is always the last highest power of 2 tested
    Dim rem_val As Double           ' rem_val will always contain the remainder not yet converted
    Dim digits_present As Integer ' 0 if no digits present yet, 1 if digits or period present

    Do
        power_val = 4294967296#   '  set maximum number allowed - must be power of 2
        digits_present = 0
        result_string = ""
        If (input_value < 0) Then result_string = "-"      ' if number is negative put in minus
        rem_val = Abs(input_value)  ' Positive and negative numbers displayed the same except for
                                    ' leading minus sign
        If (rem_val >= power_val) Then
            result_string = "Error: Absolute value of number must be less than " & power_val
            Exit Do
        End If
        Do ' ------ outer loop to get next binary character ------
            If ((power_val <= 1) And (rem_val = 0)) Then Exit Do
            If (power_val = 1) Then ' if got to end of integral portion put in leading zero and period
                If (digits_present = 0) Then result_string = result_string & "0"      ' add leading zero
                result_string = result_string & "."    ' add the period or comma to the string
                digits_present = 1
            End If
            power_val = power_val / 2        ' divide power_val by 2 to get next lower power of 2
            If (rem_val >= power_val) Then            ' if remainder greater than power_val then
                result_string = result_string & "1"    ' put a 1 in the next position in the string
                rem_val = rem_val - power_val         ' and reduce rem_val by power_val
                digits_present = 1
            Else
                ' otherwise put in a 0 if not dealing with leading zeroes
                If (digits_present = 1) Then result_string = result_string & "0"
            End If
        Loop
        If (digits_present = 0) Then result_string = result_string & "0"
        result_string = result_string & "B" ' add optional trailing B to indicate a binary number
        Exit Do
    Loop
    double_to_binary = result_string ' return result     ' return result to calling routine
End Function
```

```
Function double_to_hex(input_value As Double) As String
'   Author:  Ian Atkinson
'   Date:  March 1, 2006
'   This function converts a floating point double precision number into a
'   character string containing the hexadecimal representation of the floating point number.
'   Limitations: Numbers greater or equal to 4294967296 not allowed
'           Number of characters in result dependent on number of bits in the mantissa of the
'           computer representation of a double precision floating point number.
'           Last digit is hex representation of rounded binary number that would be produced
'           by double_to_binary routine above.

  Dim result_string As String    ' result_string contains the binary chars so far in the result
  Dim power_val As Double         ' This value is always the last highest power of 2 tested
  Dim rem_val As Double           ' rem_val will always contain the remainder not yet converted
  Dim digits_present As Integer   ' 0 if no digits present yet, 1 if digits or period present
  Dim hex_value As Integer        ' stores value of up to last four bits processed
  Dim hex_bits As Integer         ' stores number of bits so far in hex_value (0 to 4)
  Dim hex_chars As String         ' used to get hexadecimal chars to put in result

  Do
     hex_chars = "0123456789ABCDEF"
     power_val = 4294967296#  '   set maximum number allowed - must be power of 16
     digits_present = 0
     result_string = ""
     If (input_value < 0) Then result_string = "-"
     rem_val = Abs(input_value)
     If (rem_val >= power_val) Then
        result_string = "Error: Absolute value of number must be less than " & power_val
        Exit Do
     End If
     hex_value = 0
     hex_bits = 0
     Do '  ------ outer loop to get next binary character ------
        If ((power_val <= 1) And (rem_val = 0)) Then   ' see if come to end of input data
           If (hex_bits > 0) Then
              While (hex_bits < 4)    ' finish up getting final hex character if so
                 hex_value = hex_value * 2
                 hex_bits = hex_bits + 1
              Wend
              result_string = result_string & Mid(hex_chars, hex_value + 1, 1)
           End If
           Exit Do
        End If

        If (power_val = 1) Then
           If (digits_present = 0) Then result_string = result_string & "0"     ' put in leading 0
           result_string = result_string & "."     ' add the period or comma to the string
           digits_present = 1
        End If

        power_val = power_val / 2
        hex_value = hex_value * 2
        If (rem_val >= power_val) Then
           rem_val = rem_val - power_val
           hex_value = hex_value + 1
        End If
```

```
          hex_bits = hex_bits + 1

       If (hex_bits = 4) Then
          If ((digits_present = 1) Or (hex_value > 0)) Then
             result_string = result_string & Mid(hex_chars, hex_value + 1, 1)
             digits_present = 1
          End If
          hex_value = 0
          hex_bits = 0
       End If
    Loop
    If (digits_present = 0) Then result_string = result_string & "0"
    result_string = result_string & "h"   ' add optional trailing h to show that number is hex
    Exit Do
  Loop
  double_to_hex = result_string
End Function
```

www.ingramcontent.com/pod-product-compliance
Lightning Source LLC
Chambersburg PA
CBHW031808190326
41518CB00006B/238